A Field Guide in Color to
BIRDS

A Field Guide in Color to

BIRDS

By Dr. Walter Černý

Illustrated by Karel Drchal

Cathay Books

Translated by Margot Schierlová
790 colour plates, 235 b/w plates and 44 pen drawings by Karel Drchal;
672 distribution maps by Zdeněk Hedánek; 64 colour photographs
by O. Danesch (1), G. Quedens (15), H. Schrempp (17), K. Schwammberger (9),
D. Zingel (22)
Graphic design by Karel Drchal

This English version published 1975 by
Cathay Books
59 Grosvenor Street, London W1

ISBN 0 904644 10 3

Printed in Czechoslovakia
3/07/02/51

CONTENTS

Bird Identification 7
Instructions for Use and Ornithological Terms 9
Topography of the Bird's Body 12
Type Determination of Birds 14
The Avian System 19
Explanations 22
Plates 23

Divers (Gaviiformes) 24
Grebes (Podicipediformes) 26
Pelicans and Allies (Pelecaniformes) 28
Herons and Allies (Ciconiiformes) 30
Ducks, Geese and Swans (Anseriformes) 38
Birds of Prey (Falconiformes) 58
Game Birds (Galliformes) 76
Rails and Allies (Gruiformes) 82
Waders (Charadriiformes) 90
Gulls and Terns (Lariformes) 114
Auks (Alciformes) 130
Pigeons and Doves (Columbiformes) 132
Owls (Strigiformes) 136
Cuckoos (Cuculiformes) 144
Nightjars (Caprimulgiformes) 144
Rollers and Allies (Coraciiformes) 144, 146
Swifts (Apodiformes) 146
Woodpeckers (Piciformes) 148
Passerines (Passeriformes) 156

Ducks in Flight 248
Geese in Flight 256

Geese and Swans in Flight 258
Birds of Prey in Flight 260
Waders in Flight 272
Gulls and Terns in Flight 284
Illustrations of Eggs and Nests 296
Descriptions of Eggs and Nests 313
Birds Nesting in Marginal Regions and a few Accidental Birds 321
Bibliography 331
Index of Common Birds' Names 333
Index of Scientific Birds' Names 339

BIRD IDENTIFICATION

Today there is a wide range of books on how to identify birds. They have various advantages and disadvantages and differ in size, presentation and price. We have tried to produce a book which will give good value for money, which is as complete as possible and which defines the essential characteristics of the different species by coloured illustrations made specially for this purpose.

The chief aim of our book is to help the reader identify the birds he encounters in the woods and fields or watches on his window-sill. The task is by no means easy, since many species are very similar in appearance, while in others the males, females and young birds wear quite different plumage and in many cases we find transitional forms between juvenile and adult plumage or between summer and winter plumage. Only an artist with a knowledge of the subject could emphasize, in the colour plates, the elementary differences between the various types of age- and season-determined plumage so that the bird-lover can recognize the actual bird and give it its proper name. If we want to identify a bird, we can employ various methods. If we study the illustrations, we shall become familiar with the form, colouring and markings of individual species. In doing so, the image of a large number of birds will remain in our memory and we shall recognize the birds again when we see them. Another way is to use the survey of the different types so as to obtain a general impression of the bird forms we are likely to encounter over a very wide area and then, when we come to observe a species, we can refer to the detailed texts for further information.

We have illustrated as many birds as possible in flight, when we can often see the characteristic markings, the pattern of the stretched wing, the colouring of the rump and the sides (which are otherwise usually invisible).

The texts accompanying the colour plates describe — always in the same order — the important identification marks (from the aspect of field observation) for the male, female and young of each species and, if necessary, for the summer and winter plumage. They also give details of the call, habitat, breeding season, nest, eggs and migratory habits.

The only subspecies treated are those which differ markedly in some way from the main species (e.g. the subspecies of the Blue-headed Wagtail, or the Carrion Crow and the Hooded Crow).

A great deal of the credit for this book must go to the talented illustrator Karel Drchal, who fulfilled the difficult task of producing lifelike illustrations, while at the same time emphasizing important features, in a really splendid manner.

Walter Černý

INSTRUCTIONS FOR USE AND ORNITHOLOGICAL TERMS

The most important part of a handbook with lifelike colour plates is always the illustrations and a detailed written description of the individual species is unnecessary. We have therefore laid stress only on the distinctive characteristics of each species in this guide and have given the differences between the male (♂), female (♀) and juvenile (juv). Young birds in their down plumage and nestlings can best be identified by the parents feeding or escorting them, as even an experienced expert would find it difficult, without comparative material, to distinguish between nestling Song Thrushes and Redwings, or between newly hatched Pochards and Ferruginous Ducks, for example. The usual technical terms employed in the brief texts are explained in the drawings on p. 13.

In many species, the male and female bird wear more or less different plumage, making it easier to tell their sex. These differences are pointed out in the short descriptions and are also depicted in the illustrations, as long as they are useful characteristics for field observations. The young birds' plumage (juvenile plumage) usually resembles the female's. Very often differences are not noticed until the birds are caught and examined in detail. Such characteristics are obviously not suitable identification marks for field ornithological studies and we have not taken them into account.

Summer plumage is the general term for the plumage worn during the breeding season. In the spring, when it is fully developed, it is known as nuptial plumage. At the end of the breeding season, when the birds moult, their appearance changes and they don their winter plumage. Many males which are brightly coloured during the summer acquire plumage so closely resembling the females' that it is impossible to tell the difference between them under field conditions.

A good pair of field-glasses is a great help for observing details. Binocular field-glasses giving 8-fold magnification are adequate for most purposes.

The breeding area is marked black in the distribution maps, while the limit of regular distribution, which is often the normal winter area, is bounded by a line. Arrows at the foot of the distribution maps indicate the direction taken by those migratory birds that usually winter south of the Sahara. The breeding area of one or two birds which winter in Europe every year lies outside the limits of the map and in that case its position in the Arctic region or northern Asia is indicated by arrows at the top of the map.

Size: It is usually not possible to tell the size of a bird's body from superficial observation in the open, since this depends primarily on a correct estimation of distance. The impression of size is further influenced by visibility factors (light, air transparency and a non-contrasting background). We therefore chose familiar birds of known build for a comparison of size, i.e. the sparrow, Blackbird, pigeon, Mallard and goose, which the observer can watch at any time and under the most diverse conditions. It is particularly difficult to estimate the distance of flying birds (e.g. birds of prey, pigeons, thrushes and crows), but, here again, practice makes perfect.

B o d y f o r m — in particular the shape of the wings and beak and the length of the tail — is important for the identification of many species of birds. In the case of swimming aquatic birds, special attention should be paid to the shape of the head, the way the tail is carried, how much of the body is submerged, etc.

In conjunction with other characteristics, especially the colouring of the plumage, the p a t t e r n shows the most important identification mark. The e y e s t r i p e is a conspicuous light or black line in the region of the eye (e.g. Nuthatch, Goldcrest, Blue-headed Wagtail), and the c r o w n s t r i p e, a similar row of feathers on the crown (e.g. Aquatic Warbler, Whimbrel). The l o r e is a coloured, often featherless stripe leading from the corner of the beak to the eye. The m o u s- t a c h i a l s t r i p e runs from each corner of the beak down the sides of the neck (e.g. Hobby, Green Woodpecker, Bearded Titmouse). The pattern on the upper surface of the wings provides pointers which are very important for the identification of flying birds. There may be only a characteristic white wing stripe on the posterior margin of the wing, while sometimes the pointers are white, dark or metallic patterns or patches (e.g. the specula of dabbling ducks). Identification marks of this kind are stressed in the text and in the illustrations. The tail may have a characteristic dark or white border and other patterns are also found on the tail feathers, e.g. cross bands, a triangular spot, etc.; in many species of birds, the marginal feathers of the tail are white and are thus particularly noticeable when the tail is spread. The rump is often coloured differently from the tail and back, and can be a very valuable identification mark, even in quite similar species (e.g. Golden and Grey Plover, Marsh and Hen Harrier).

The f a c i a l d i s c is the term applied to the radially organized plumage round the beak of a few birds (owls, harriers). It frames their face, in which the eyes are directed forwards.

The observer must therefore quickly note as many of these characteristics as possible to be able to identify a bird with certainty.

B i r d s' v o i c e s (and especially their song) are one of the most important field characteristics and a knowledge of them greatly simplifies the identification of a species. In the case of species which conceal themselves or closely resemble others in colouring and form, the voice is often the only means of quick identification. Take the outward similarity of certain reed warblers, leaf warblers and tree-creepers, for instance. The highly secretive rails and birds active only after dusk (owls, nightjars) can hardly be detected by optical means. Knowledge of the voices of such birds is one of the field ornithologist's most important aids in such cases. All we have been able to do is to represent the birds' commonest calls phonetically, but the reader should also have access to good quality records of their voices, especially their songs.

The way birds m o v e often helps in their identification. Many of them walk, hop, waddle or run in a manner peculiar to the species. Some can climb vertical tree trunks (woodpeckers, nuthatches, tree-creepers). Many aquatic birds dive, the technique and the time they spend under water varying with the species. Very often, birds perform characteristic movements, e.g. they may vibrate, jerk, twirl or cock their tail, wriggle, rock their body or curtsey (Godwit, Little Owl) or jerk their head (Coot, pigeons and doves). Their flight may be accompanied by an audible sound, e.g. whirring (Partridge), or clattering (Pheasant, Capercaillie) and sometimes similar species can be differentiated by the rate at which they beat their wings (gulls, Goshawk and Sparrow Hawk). Some birds hover in a characteristic manner (Kestrel, Common Tern, Great Grey Shrike). Even the path of their flight (straight, undulating or in 'hops') displays certain characteristic features. A number of birds have a typical form of flight when courting (Skylark, Tree Pipit). There are thus clearly plenty of features by which birds can be identified, but the observer must, in the first place, learn from his own experience.

A knowledge of nests and eggs will give the field ornithologist some useful supplementary hints and we have therefore illustrated a few clutches and nests. It is sometimes necessary to look for a nest to obtain exact proof of a bird's breeding habits, but one should do so only when absolutely necessary, so as not to disturb the birds. In particular, one should avoid handling the eggs.

The breeding season, in the broadest meaning of the term, comprises the whole of the mating and breeding process, i.e. including courtship (which may start a whole month before the eggs are laid), nest-building, egg-laying, the incubation period, the hatching period and the period of post-natal care (until the young are independent). For practical reasons we have described only the time from the commencement of egg-laying to the point where the parents stop caring for the young as the 'breeding season'. We were unable to take into account differences between the start of the breeding season in the most southerly and northerly latitudes of the European breeding area (which are sometimes as much as 6 weeks), or the prolonged training period of geese, a number of ducks and game birds, during which the family remains united for a comparatively long time. The breeding season is also prolonged if the normal brood is lost and a second clutch is laid. The breeding season times given in this book apply mainly to central and western Europe. In southern Europe breeding starts 2—3 weeks earlier and in northern Europe 2—3 weeks later. Unless explicitly stated otherwise, there is only one brood a year.

The size of the clutch is given as the average number of eggs in a complete clutch. The actual number may be smaller, especially in late clutches, or it may be abnormally large, e.g. if two females lay their eggs in the same nest as often occurs with ducks.

Similar big differences are found in the migration time (up to 10 weeks in widespread species whose range stretches from the most southerly parts of Europe to the Arctic circle). We therefore again considered it best to base ourselves on conditions in central Europe. One species of bird may be found the whole year round in a given observation area. This means that it is either a resident bird, which stays near its breeding site the whole time (e.g. House Sparrow, Nuthatch, Marsh Tit), or a partial migrant, i.e. when only part of the local population migrates, while the rest are joined by birds of the same species from somewhere else. Many species are partial migrants in central Europe, but behave as full migrants in northern Europe and as resident birds in southern and western Europe. Migrants are birds which leave their breeding place every year at roughly the same time, migrate to a distant region for the winter and regularly return again to their breeding place in the spring. Such birds can be encountered in our observation area during the summer only (summer visitors). Typical migrants are to be found among birds which live on insects (swallows, swifts, cuckoos) and among birds which seek their nourishment in or on water (terns, storks). Every year, their migration takes them on journeys covering several thousand miles. Various intermediate forms between the above categories also exist, e.g. nomadic birds, which likewise leave their breeding place, but roam about during the winter, never too far from 'home'. Irruptive birds are another type. These are species with irregular migration habits, which usually appear in large numbers at intervals of several years. Irruptions are usually caused by birds living in northern and eastern Europe (e.g. Redpoll, Crossbill, Waxwing, Thin-billed Nutcracker, Red-footed Falcon). In an observation area lying between the breeding and winter area of a given bird, the species appears only during migration (birds of passage). Species which breed in northern Europe and the Arctic region appear in the rest of Europe as winter visitors. Birds which stray from their normal areas and migration routes, or are driven off course by storms, and appear in regions to which they are foreign, are known as accidental birds (e.g. marine birds found inland, or a few species from Asia and North America — see p. 321).

TOPOGRAPHY OF THE BIRD'S BODY

A. Kestrel

B. Wing of Mallard (drake)

1. Crown
2. Forehead
3. Beak with cere
4. Throat
5. Side of neck
6. Moustachial stripe
7. Breast
8. Bastard wing
9. Flanks
10. Belly
11. Tibial feathers
12. Tarsus
13. Toes
14. Under tail-coverts
15. Upper tail-coverts
16. Tail feathers
17. Orbital ring
18. Nape
19. Cheek or ear coverts
20. Carpal joint
21. Mantle
22. Back
23. Rump
24. Primaries
25. Tail band
26. Wing coverts
27. Secondaries
28. Speculum
29. Scapulars

A.

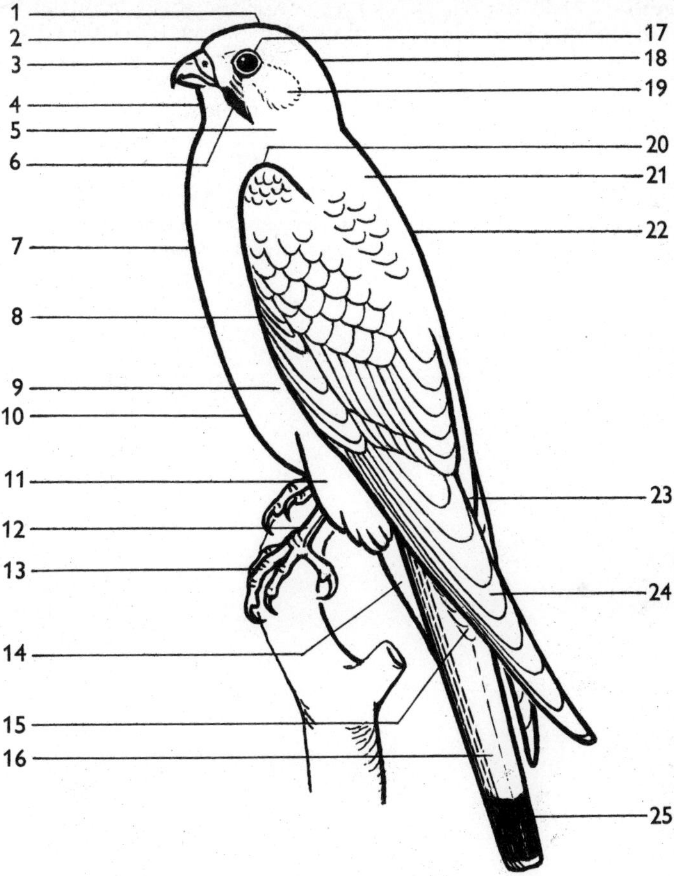
1
2
3
4
5
6
7
8
9
10
11
12
13
14
15
16
17
18
19
20
21
22
23
24
25

B.

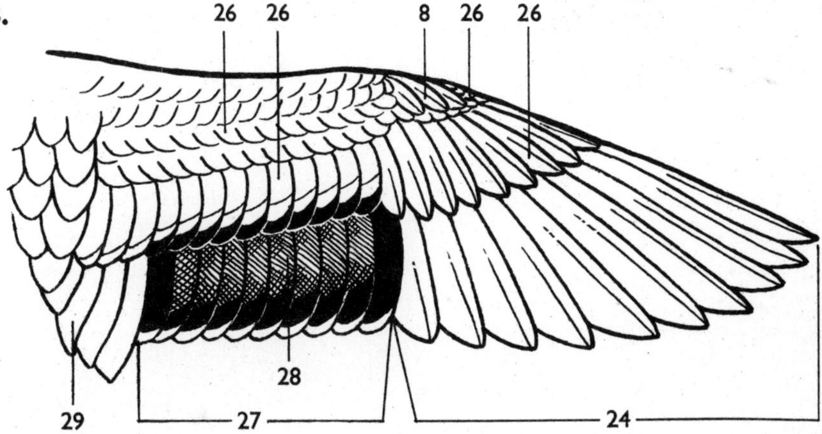
26
26
8
26
26
28
29
27
24

TYPE DETERMINATION OF BIRDS

In most cases, even the beginner will not find it too difficult to determine the group to which a given bird belongs. It is easy enough to distinguish between a songbird, an owl and a duck, but an inexperienced observer may not be able to tell a wryneck (one of the woodpeckers) from a songbird, a flying cuckoo from a bird of prey, or a diver from a duck. That is only at first, however. There are so many differences in the form of the various parts of the body, in the way birds move and sit, and in their behaviour in general, including vocal expression, that determination of the group (order, sub-order and sometimes even family) to which a given bird belongs can hardly go wrong.

The main types of birds are illustrated on the following pages.

Aquatic and Littoral Birds

are completely dependent on water (or at least for their food)

		Similar to
Divers and Grebes Lie deep in water, dive often and long. Wide body, long neck, sharp-pointed beak, no tail.	 Great Crested Grebe	 Duck
Pelicans and Allies Large, longish neck, hook-tipped beak. Dive.	 Cormorant	 Diver

Herons and Allies

Grey Heron

Crane

Large. Long neck and legs. Long beak tapering to point.

Ducks, Geese and Swans

Duck

Goose

Coot

Longish neck, fairly large head with laterally directed eyes. Bill mostly flat, with 'nail' at tip. Short tail. Wings often have brightly coloured speculum. Include many skilled divers.

Rails and Allies

Coot

Crane

Stork

Water Rail

Sandpiper

Large, with long legs and neck, or duck to thrush size.

Aquatic and Littoral Birds	Similar to

Waders

Small to moderately large, fairly large head, long legs. Long, thin beak, short tail. Narrow, curved wings.

Lapwing

Sandpiper

Water Rail

Gulls and Terns

Moderately large to large, with long, narrow wings, short legs. Skilled fliers.

Black-headed Gull

Auks

Fairly large head, flat-sided beak, short legs and tail. Moderately large. Dive extremely well. Strictly marine birds.

Auk

Duck

Diver

Tree- and Ground-dwelling Birds

Types seen mainly in the air are depicted in flight.

Birds of Prey

Hooked beak. Mostly large birds with long wings. Fly fast and skilfully, often glide in circles.

Kestrel

White-tailed Eagle

Game Birds

Ground-dwellers, mostly of moderate size, with powerful feet for scratching. Small head, thick beak with curved tip.

Partridge

Pheasant

Pigeons and Doves

Small head, tripping gait, nodding head.

Wood Pigeon

Similar to

Cuckoos

Slender. Long, gradated tail. Short legs.

Cuckoo

Sparrow Hawk

Owls

Large head and eyes. Eyes directed forwards. Round-tipped wings, almost inaudible flight.

Tawny Owl

Nightjars

Large head with large eyes. Fly similarly to swallows.

Nightjar

Kestrel

Swifts

Resemble swallows in flight. Curved wings.

Swift

Swallow

Rollers and Allies

Sparrow to jay size. Colouring usually bright and metallic. Fairly long beak.

Kingfisher

Hoopoe

Blackbird

Woodpeckers

Climb vertical tree trunks. Wedge-shaped beak. Tail used as support.

Green Woodpecker

Passerines

Small to moderately large, with mostly short beak and longish tail.

Crow

Swallow

Warbler

Chaffinch

Wagtail

THE AVIAN SYSTEM

Our book follows the scientific system which classifies the large avian orders in groups. The orders in turn are subdivided into families, the families into genera and the genera into species. In the scientific names, the first word stands for the genus and the second for the species. Those orders and families mentioned only under Accidental Birds — p. 321 — are marked below with an asterisk:

Order	Family
Divers (Gaviiformes)	Divers (Gaviidae)
Grebes (Podicipediformes)	Grebes (Podicipedidae)
Tubenoses (Procellariiformes)*	Storm Petrels (Hydrobatidae)* Shearwaters (Procellariidae)*
Pelicans and Allies (Pelecaniformes)	Cormorants (Phalacrocoracidae) Gannets (Sulidae)* Pelicans (Pelecanidae)*
Herons and Allies (Ciconiiformes)	Herons (Ardeidae) Storks (Ciconiidae) Ibises (Threskiornithidae)
Flamingos (Phoenicopteriformes)*	Flamingos (Phoenicopteridae)*
Ducks, Geese and Swans (Anseriformes)	Ducks (Anatidae)
Birds of Prey (Falconiformes)	Vultures, Eagles, Buzzards, etc. (Accipitridae) Falcons (Falconidae) Ospreys (Pandionidae)

Order	Families
Game Birds (Galliformes)	Grouse (Tetraonidae) Pheasants, Partridges and Quails (Phasianidae)
Rails and Allies (Gruiformes)	Crakes, Rails and Coots (Rallidae) Cranes (Gruidae) Bustards (Otididae)
Waders (Charadriiformes)	Oystercatchers (Haematopodidae) Plovers and Lapwings (Charadriidae) Snipes and Sandpipers (Scolopacidae) Phalaropes (Phalaropidae) Avocets and Curlews (Recurvirostridae) Stone Curlews (Burhinidae) Pratincoles (Glareolidae)
Gulls and Terns (Lariformes)	Gulls (Laridae) Terns (Sternidae) Skuas (Stercoraridae)
Auks (Alciformes)	Auks (Alcidae)
Pigeons and Doves (Columbiformes)	Pigeons and Doves (Columbidae)
Sandgrouse (Pterocliformes)*	Sandgrouse (Pteroclidae)*
Owls (Strigiformes)	Owls (Strigidae) Barn Owls (Tytonidae)
Cuckoos (Cuculiformes)	Cuckoos (Cuculidae)
Nightjars (Caprimulgiformes)	Nightjars (Caprimulgidae)
Swifts (Apodiformes)	Swifts (Apodidae)
Rollers and Allies (Coraciiformes)	Rollers (Coraciidae) Kingfishers (Alcedinidae) Bee-eaters (Meropidae) Hoopoes (Upupidae)

Woodpeckers (Piciformes)

Woodpeckers (Picidae)

Passerines (Passeriformes)

Swallows (Hirundinidae)
Waxwings (Bombycillidae)
Larks (Alaudidae)
Wagtails (Motacillidae)
Dippers (Cinclidae)
Wrens (Troglodytidae)
Shrikes (Laniidae)
Accentors (Prunellidae)
Warblers (Sylviidae)
Goldcrests (Regulidae)
Flycatchers (Muscicapidae)
Thrushes (Turdidae)
Bearded Tits (Paradoxornithidae)
Long-tailed Tits (Aegithalidae)
Penduline Tits (Remizidae)
Titmice (Paridae)
Nuthatches (Sittidae)
Tree-creepers (Certhiidae)
Buntings (Emberizidae)
Finches (Fringillidae)
Weavers (Ploceidae)
Starlings (Sturnidae)
Orioles (Oriolidae)
Crows (Corvidae)

EXPLANATIONS

The following customary symbols and terminological abbreviations are used in the plates:

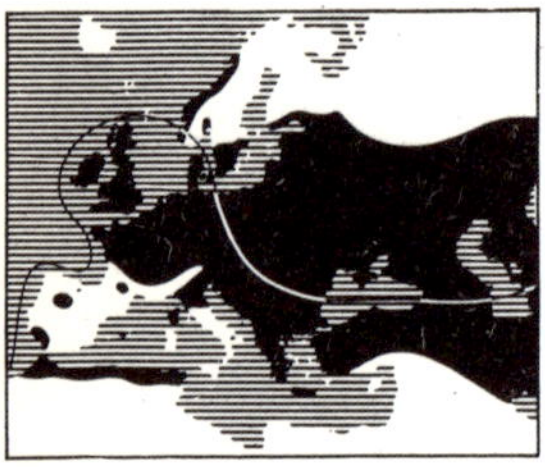

♂ = male
♀ = female
D = downy plumage
juv = juvenile plumage
imm = immature (plumage not fully coloured)
ad = adult plumage
S = summer (nuptial) plumage
W = winter plumage
LV = light variant
DV = dark variant

In the European distribution maps, the b r e e d i n g a r e a is marked black. The regular w i n t e r a r e a is indicated by a line which usually marks the northern margin of the winter quarters and sometimes outlines the most important winter areas. The winter areas often lie outside the limits of the map (usually to the south) and so, occasionally, do the breeding areas. In both cases their localization is indicated by arrows. In the American distribution maps the entire range of the species (excluding 'passage' areas) is filled in black.

The European maps on p. 24 (left) and p. 246 (middle) each contain distribution data for two species. In the former, the distribution of the White-billed Diver is marked by the dot-and-dash line and in the latter, the distribution areas of the Carrion Crow and the Hooded Crow are shown side by side, in different shading.

The Roman numerals in the text stand for the months of the year.

The headings include both English and American common names where different.

PLATES

1 **GREAT NORTHERN DIVER - COMMON LOON** *(Gavia immer)* Size of goose. In summer plumage distinguishable from Black-throated Diver by size of head and massive black beak. ♂ = ♀. In winter and juvenile plumage cannot always be reliably distinguished from Black-throated Diver, but thicker beak, thicker neck and squarer head sometimes help. Voice: at breeding time a tremulous, carrying wail, in winter mostly silent. Arctic species; nests with Black-throated Diver on lakes. Rare. Breeding season: VI—VIII. Diet: fish. Rare winter bird, appearing regularly only near North Sea coast (IX—V). Inland accidental bird. **White-billed Diver** *(Gavia adamsii) (1 a)*: rare winter bird on European coasts and inland accidental bird. Inhabits coasts of arctic Russia. Distinguishable from Great Northern Diver by light, ivory-coloured beak with straight upper mandible and upcurving lower mandible.

2 **BLACK-THROATED DIVER - ARCTIC LOON** *(Gavia arctica)* Almost size of goose, tailless. Much of body submerged, masking actual size. ♂ = ♀. Winter plumage like other divers', but differences in shape of beak and size. Indistinct scaly spots on back and often spots or ± dark shading on neck (see other species). Fast wing beat, no specula. Exclusively aquatic, dives often and very easily, normally remaining up to 1 min. under water. Voice: barking single calls and a longer wailing call at breeding time. Nests beside lakes with good fish supply. During migration and in winter seen chiefly on coast and large lakes. Not common. Breeding season: V—VI. Nest built right beside water; 2 brown, sparsely spotted eggs. Young fed and led 2 months by both parents. Diet: fish. Winter bird from X to III.

3 **RED-THROATED DIVER - RED-THROATED LOON** *(Gavia stellata)* Usually smaller than Black-throated Diver, distinguishable by slender beak with upcurving lower mandible. Easily recognized in summer plumage by red patch on neck and absence of pattern on back. ♂ = ♀. In winter and juvenile plumage generally lighter than Black-throated Diver; more white, especially on head, otherwise apparently up-tilted beak main criterion. Breeds mainly on coast, less frequently on lakes. Comparatively rare. Same habits as Black-throated Diver, but sometimes nests in colonies. Voice: at breeding time cackling or long series of loud calls, also plaintive mewing. Winter birds usually remain until breeding season. 2 olive green to brown eggs with a few black-brown spots laid in V—VII. Diet: fish. Bird of passage and winter bird; migration X—XII and III—IV.

1

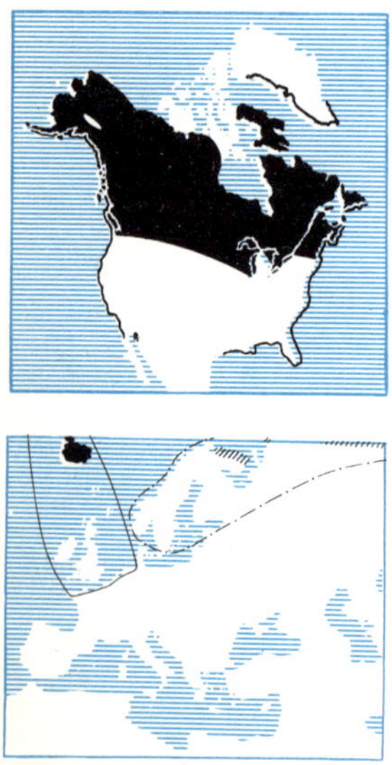

2

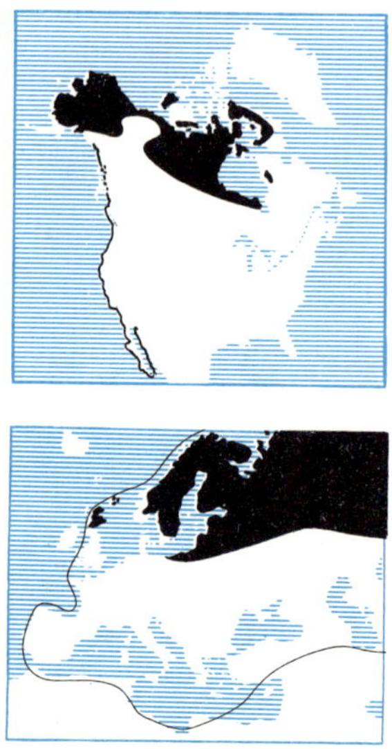

3

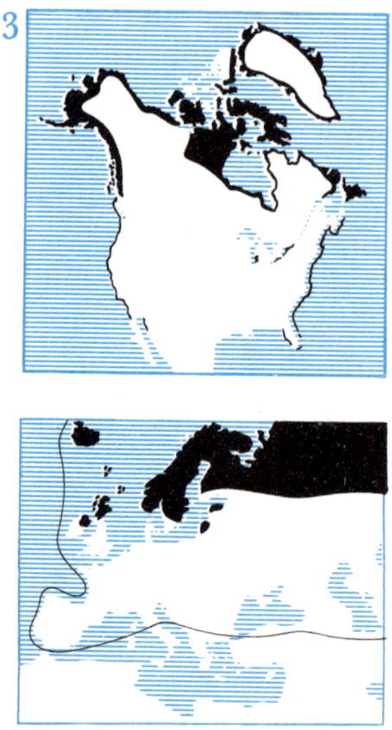

1 W
1 S
1 a W
2 W
2 S
2 D
3 W
3 S

Order: Grebes - Podicipediformes **Family: Grebes -** Podicipedidae

1 **GREAT CRESTED GREBE** *(Podiceps cristatus)* About size of duck. Pure white neck, sharp beak, glossy white plumage on breast. Forked crest and expansible ruff on head at mating time. Tailless. ♂ = ♀. Downy young have black longitudinal stripes, juveniles dark stripes on cheeks and neck. Lives entirely on water, dives frequently. Courtship: breast to breast with erect body and extended neck, shakes head, utters loud calls ('greck' and 'kek kek kek', also barking 'gorr'). Breeding season: IV—VIII. Nest made of aquatic plants, floats on water, anchored to edge of reed bed. 3—6 eggs, first white, later brown, covered when bird leaves nest. Newly hatched young carried by parents in warm dorsal plumage when swimming. Main diet: fish. Partial migrant (VIII—XI), often winters on ice-free lakes and rivers and off coast.

2 **RED-NECKED GREBE** *(Podiceps grisegena)* Smaller and more robust than Great Crested Grebe. In summer plumage, short, thick neck russet red, sides of head grey, short 'horned' crest, base of beak yellow. Winter plumage: grey neck, 'horns' barely detectable. ♂ = ♀. Downy young resemble those of Great Crested Grebe. Entirely aquatic, dives often. In breeding season frequents low-lying ponds and lakes with dense vegetation. Likes shelter. Courtship as Great Crested Grebe. Call ('ga' or 'kek') uttered frequently in series, in spring often loud, like crying foal, audible for long distances. Comparatively rare bird. Breeding season IV—VII. Same type of nest as Great Crested Grebe. 3—5 eggs, first white, later brown. Downy young carried on parents' back. Diet: small fish, frogs, insects, molluscs. Partial migrant (III—IV and IX—X). Winters solitarily.

3 **LITTLE GREBE** *(Podiceps ruficollis)* Smallest grebe, thickset body, short neck. In summer distinguishable by chestnut neck and gleaming greenish yellow corners of beak. In winter mainly rusty brown. ♂ = ♀. No white specula on wings. Downy young almost black, with distinct stripes down body. Always stays on water. Swims mostly with ruffled feathers, body relatively high out of water; dives frequently. Common species on marshy ponds and lakes, in winter on stagnant and running water with no vegetation. Courting birds utter prolonged, explosive trills ('bibibibee'), swim towards each other at top speed and dive. Breeding season: IV—VII, two broods a year. Floating nest with 5—6 eggs, first white, later brown. Diet: chiefly insects, less often molluscs, in winter also fish. Migration VIII—XII, according to weather. Many birds remain in winter quarters until III.

1

2

3

1 W
1 S
1 D
2 S
2 W
3 S
3 W

Order: Grebes - Podicipediformes **Family: Grebes** - Podicipedidae

1 **BLACK-NECKED GREBE - EARED GREBE** *(Podiceps nigricollis)* About size of partridge, with thin, apparently tilted beak and steeply sloping frontal crest. In nuptial plumage has black neck and flat, rusty yellow lateral head tufts. ♂ = ♀. In winter differs from very similar Slavonian Grebe by shape of beak and head and blurred division between black crown and white neck. Downy young dark, with indistinct stripes. Entirely aquatic, dives frequently. Common species on overgrown ponds and lakes. Breeding season V—VII, one brood a year. Nests in colonies, usually together with Black-headed Gulls. Voice: a whistling 'poo-eep'. Nest inconspicuous heap of aquatic plants. 3—4 eggs, first white, later brown. Downy young carried on parents' back. Diet: chiefly water insects. Migrant, migrates in IV and IX—XI.

2 **SLAVONIAN GREBE - HORNED GREBE** *(Podiceps auritus)* Size of partridge, distinguishable from Black-necked Grebe by orange yellow, expansible tuft of feathers on side of head and russet neck. In winter resembles Black-necked Grebe, but has straight, light-tipped beak, flatter crown and sharp division between black head and white cheeks. ♂ = ♀. Striking white specula seen on wings during flight. Downy plumage blackish, with distinct stripes down back. Entirely aquatic, likes diving. Inhabits freshwater lakes, in winter coasts, appears regularly in small numbers on large, ice-free inland stretches of water. Breeding season V—VII. Courting behaviour and nest like Black-necked Grebe, also often nests in colonies; 3—5 eggs. Trilling mating call, otherwise almost silent. Diet: mainly water insects and fish. Winter bird from IX to III, occasionally to V.

Order: Pelicans and Allies - Pelecaniformes **Family: Cormorants** - Phalacrocoracidae

3 **CORMORANT - GREAT CORMORANT** *(Phalacrocorax carbo)* About size of goose. Black, with slender, hook-tipped beak and bare throat. Summer plumage black, sides of head and neck mostly white. ♂ = ♀. Juvenile brown on back and ± white on underside. Stays mainly on water, swims largely submerged, with raised head and beak. Dives frequently. Rests erect, often with spread wings, on stone, post or tree protruding from water. In flight, with stretched head and neck and longish, stiff tail, forms cross. Voice: harsh croaks and grunts. Inhabits coastal water and mouths of rivers; when migrating appears in regions with fish-ponds. Nests gregariously on tall deciduous trees. Breeding season: IV—VII. Nest made of twigs and sticks; usually 3—4 light blue eggs covered with chalky layer. Diet: fish. Resident and nomadic bird.

1

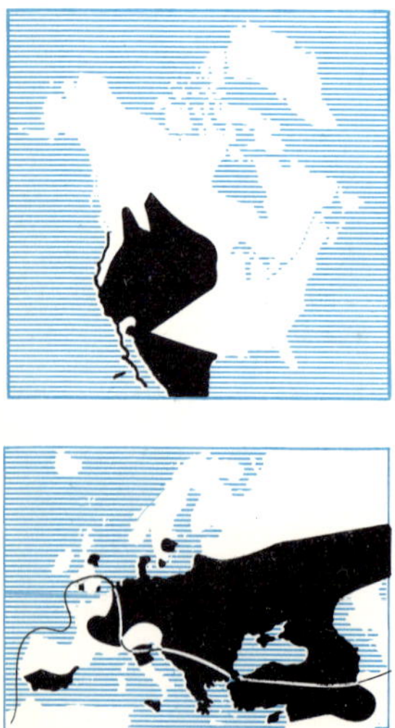

2

3

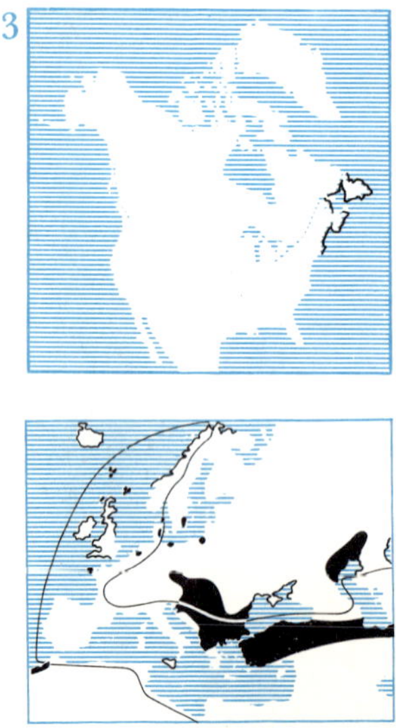

1 W
1 S
2 S
1 D
2 W
3 juv
3 ad

1 **GREY HERON** *(Ardea cinerea)* Similar size and shape as stork. Grey back, black eye stripe terminating in long plumes. ♂ = ♀. Juvenile has no eye stripe. Downy young have long, silvery down, forming long crest on head. Flies with slow wing beats and retracted head, with neck folded in 'S'. Voice: always utters loud, harsh 'frarnk' when flying, other (mostly croaking) sounds on nest. Chiefly frequents rivers and regions with ponds and lakes. Common bird. Breeding season: III—VII, nests in colonies. Builds largish nest of twigs in tree-tops. Eggs: 4—5 light bluish green. Diet: mainly fish, caught in typical 'lurking' position, also amphibians and fieldmice. Partial migrant, starts migrating in IX. Many birds do not migrate.

2 **PURPLE HERON** *(Ardea purpurea)* Slightly smaller than Grey Heron, with slimmer neck. Mainly reddish brown. Juvenile has completely reddish brown back. ♂ = ♀. Like Grey Heron in flight, but distinguishable by colouring and longer toes. Voice: similar to Grey Heron's, but softer, call more like 'reh-ehb'; utters croaking and grunting sounds on nest. Sporadic breeder in extensive reed and sedge beds. Breeding season IV—VII; nests in colonies. Fairly large nest made of rush stems, rests on bent sedge or reeds just over water. Eggs: 4—6, bluish green. Young nidicolous, but leave nest if disturbed. Nesting period 2 months. Diet: mainly fish, to lesser degree amphibians, insects, molluscs. Migrant. Young birds disperse in VII, migration in VIII and IV.

3 **BITTERN** *(Botaurus stellaris)* Large, robust member of heron family with loose plumage, short legs and thick neck. ♂ = ♀ = juv. Downy plumage reddish brown. Lives secretively in dense reed-banks by large ponds and lakes and in swampy areas. Well protected by colour and markings. If alarmed, stands rigid like post. Identifiable in flight by size, retracted head, markings and rounded, arched wings. Voice especially audible at night. Mating call a deep, booming, carrying 'ee woomp', uttered several times. Comparatively rare bird. Breeding season: IV—VII. Nests solitarily. Nest made of sedge and reed blades in rushes, just over shallow water. Eggs: usually 5—6, olive brown. Young nidicolous, fed about one month in nest, then disperse into reeds; fledged at 8 weeks. Partial migrant. Migration: II—IV and VIII—X.

1

2

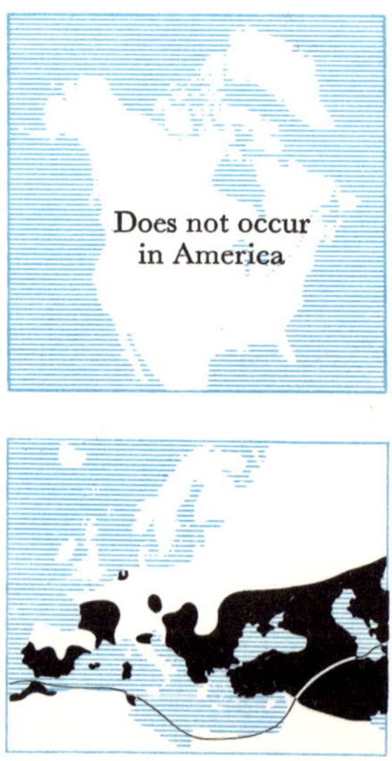

3

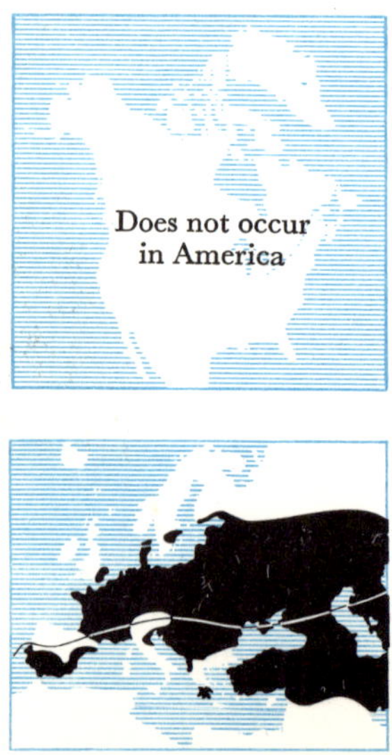

1 juv
1 ad
1
2
3
3
2 ad

1 **LITTLE EGRET** *(Egretta garzetta)* Considerably smaller than Grey Heron. Snow-white, with black beak, black legs and yellow feet. Distinguished from Great White Egret and Spoonbill (also white) by smaller size and shape of beak and in flight by quicker wing beat (from Great White Egret) and heron-like contours (from Spoonbill). In nuptial plumage wears long plumes on back and has long feathers on head and nape. ♂ = ♀. Inhabits swampy and flood-prone areas in warm lowlands. Nests in colonies, usually together with other herons. Nest made of twigs, with distinct bowl, but peripherally almost transparent; built in tree-tops or in dense bushes. Eggs: 3—5, light greenish blue, laid in V—VII. Diet: fish, amphibians, worms, insects, which are often chased in shallow water. Migrant. Migration: III—V and IX—XI.

2 **GREAT WHITE EGRET - COMMON EGRET** *(Egretta alba)* Size of Grey Heron, pure white. Much larger than Little Egret and has no long head feathers. Distinguished from Spoonbill chiefly by shape of beak. At mating time has long dorsal plumes, which are soon lost. ♂ = ♀. Winter plumage = juvenile plumage. Downy young silvery white, with 'bonnet' on head. Resembles heron in flight. Voice: croaking sounds, only on nest. Silent flight. Inhabits open lakes, ponds and unspoilt river banks. Nests hidden in reedbanks, usually in colonies. Often stands in open on sandbanks or in shallow water at margin of sedge, also perches on trees. Rare. Breeding season: IV—VII. Nest like Purple Heron's, built of sedge and rushes near surface of water. Eggs: 3—5, light blue. Young nidicolous, fledged at about 6 weeks. Diet: chiefly fish. Nomadic and migratory bird. Migration: III and IX—XI, seldom remains over winter.

3 **SQUACCO HERON** *(Ardeola ralloides)* Almost size of crow, robust. Mainly ochre yellow, with white wings, rather thick neck and ragged plumage. ♂ = ♀. Winter plumage (which closely resembles juvenile plumage) lacks long head plumes. Distinguishable in flight from Little Bittern by light back and white rump and from Little Egret (with which it can easily be confused at a distance because of its light colouring) by usually ochre brown back. Voice: hoarse croak, only on nest. Lives secretively in swamps and flood-prone regions in warm lowlands. Rare. Breeding season: V—VII. Usually nests together with other herons. Nest built of brushwood, sometimes rush stems, in bushes. Eggs: 4—6, bluish. Young cared for in nest at least one month. Diet: chiefly insects, also frogs and fish. Migrant, some birds nomadic. Migration: IV—V and VIII—IX.

1

2

3

1
1 S
2 S
3
3 juv
3 ad

Order: Herons and Allies - Ciconiiformes **Family: Herons -** Ardeidae

1 **NIGHT HERON - BLACK-CROWNED NIGHT HERON** *(Nycticorax nycticorax)* Size of large crow, stocky, with short legs, thick neck, large head and thick beak. In summer has 2—4 band-like, erectile head plumes. Winter plumage has none. ♂ = ♀. Juvenile plumage brown, with light spots like droplets. Downy plumage long and fawn brown. In flight shows characteristically rounded head and wings. Flight also always accompanied by loud, harsh 'guok' calls. Rather rare nester in extensive swamps and pond regions, also in grassy woods beside large rivers. Flies in search of food after dusk. Nests in colonies. Breeding season: IV—VII. Nest: small, like basket, made of twigs and brushwood in bushes and on trees. Eggs: 3—5, bluish green. Young completely fledged at 7—8 weeks. Diet: chiefly fish and frogs, also insects. Migrant. Migration: IV—V and VIII—X.

2 **LITTLE BITTERN** *(Ixobrychus minutus)* Smallest heron, about size of jay, ♂ black and light ochre yellow, ♀ brown and ochre yellow, with stripes down neck and breast. Juvenile plumage similar, but more boldly striped. Downy plumage ochre yellow, short and thick. Recognizable in flight by small size. Voice: mating call of ♂ regular, often long-repeated 'hogh'. Common nester in vegetation belt beside all inland waters. Hides in reeds and adopts 'post' position if in danger; most often seen when flying. Breeding season: V—VII. Nest, conical basket of twigs and reed, rush or reed-grass stalks, built among sedge or in bushes close to water level. Eggs: 4—6, chalky white. Occasionally two broods in a year. Diet: fish, amphibians, insects, molluscs. Migrant. Migration: IV and VIII—IX.

Order: Herons and Allies - Ciconiiformes **Family: Ibises -** Threskiornithidae

3 **GLOSSY IBIS** *(Plegadis falcinellus)* About size of crow, with rather thin, curved beak. Apparently almost black plumage actually mainly reddish brown with metallic lustre. ♂ = ♀. Juvenile plumage dull and dark brown. Downy plumage dull black. Flies with outstretched neck and legs, has wide, round-tipped wings and fast wing beat. Also perches on trees. Voice: mostly silent, grunting sounds on nest. Rare. Inhabits swampy flood areas, river mouths and steppe lakes. Occurs singly as accidental bird in fishpond regions. Nests in colonies. Breeding season V—VII. Nest built of sedge blades or brushwood, low in bushes. Eggs: 3—4, bluish green. Young leave nest at about 14 days, form flocks and are fed together. Diet: insects, worms, molluscs, etc., gathered in mud. Migrant. Migration: IV and VIII—X.

1

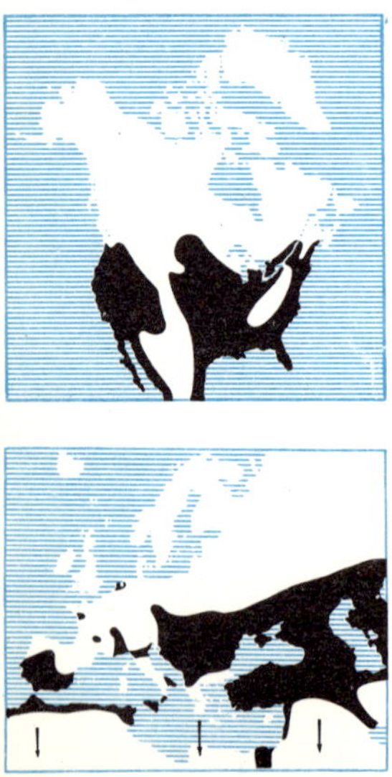

2

3

1 juv
1 ad
1
2 ♂
2 ♀
2 ♂
3
3 ad

1 **BLACK STORK** *(Ciconia nigra)* Same size as White Stork. Mainly black, with metallic lustre, white only on belly. ♂ = ♀. Juvenile plumage dull brown, with grey-green beak and legs. Downy plumage like White Stork's. Behaves like White Stork, but clatters beak less often and utters whistling and spitting sounds. Rare and timid, chiefly inhabits mountain forests and wooded hills with streams. Secretive bird. Solitary nester. Breeding season: IV—VIII. Nest built of branches and twigs near top of old tree, rarely on rocks, usually hidden in forest. Eggs: 3—5, white with greenish tinge. Incubation period one month, hatching period at least 2 months, then 2 more weeks feeding and being looked after. Diet: fish, frogs, aquatic insects caught mostly in mountain streams. Migrant. Migration: III—IV and VIII—IX.

2 **WHITE STORK** *(Ciconia ciconia)* Large bird, mainly white, with bright red beak and legs. Head and neck extended during flight (unlike herons); legs stretch beyond 'lopped' tail. ♂ = ♀. Juveniles have reddish brown beak and legs. Downy plumage white and furry. Adults clatter beak with great rapidity, throwing their head back; otherwise voiceless. Inhabits damp, marshy meadow country in fishpond regions and riverside woods. Breeding season: IV—VII. Large nest, made of sticks and twigs, usually built on roof-top, disused factory chimney or tree. Eggs: 3—4, white. Young cared for about 2 months in nest, then escorted for a time by parents. Diet: frogs, fish, snakes, insects, small mammals. Migrant. Migration: II—IV and VIII—IX. In exceptional cases does not migrate.

3 **SPOONBILL** *(Platalea leucorodia)* Smaller than stork. White, with long, spatulate bill. At mating time wears crest of long, narrow feathers. ♂ = ♀. Winter and juvenile plumage lack long head feathers. Downy plumage white. Flies with outstretched head and legs. Looks for food in shallow water, with swinging head and chattering beak. Voice: mostly silent. Rare nester in a few parts of central and southern Europe, occurs in sedge swamps with bushes and rushes, also in swampy river mouths. Nests in colonies. Breeding season: IV—VII. Nest, made of old sedge and rush blades, usually stands on bent sedge; low herons' nests also occasionally accepted. Eggs: 3—5, white. Young cared for about 2 months. Diet: aquatic insects, crustaceans, worms, molluscs, fish. Migrant. Migration: IV—V and VIII—IX, rare attempts to remain over winter.

1

2

3

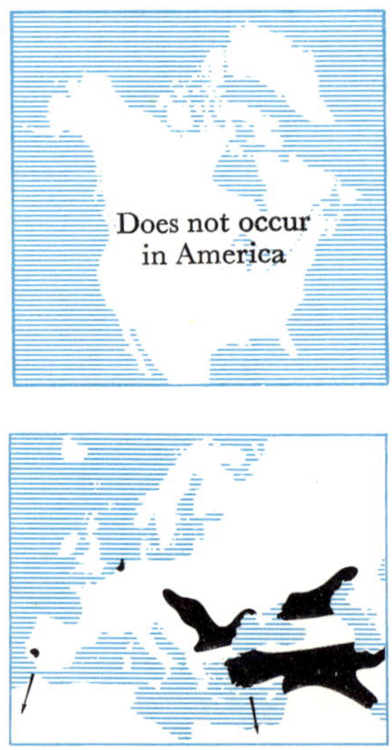

1 ad
1 juv
3 W
2 ad
3 S

1 **MALLARD** *(Anas platyrhynchos)* Largest dabbling-duck. ♂ in nuptial plumage unmistakable, ♀ brown-speckled. Orange red legs, blue, white-rimmed speculum on wings. ♂ in winter plumage resembles ♀. Juvenile plumage like ♀. Downy plumage olive brown with yellow spots and dark eye stripe. Flies well, with quick wing strokes, making high, rhythmical sound. Voice: ♂ deep 'quek ek', when courting high-pitched 'weep'; ♀ 'quarkquark quak'. Widespread, common species, especially inland on stagnant water and in winter on ice-free rivers and coasts. Breeding season: III—IV. Nest on ground under bush, usually near water, made of dry plant parts, padded with black and white down after egg-laying. Eggs: 7—11, green or blue-green, incubated about 28 days. Diet: chiefly vegetarian. Resident and migratory bird. Migration: mainly II—IV and VII—XI.

2 **GADWALL** *(Anas strepera)* Slightly smaller than Mallard. ♂ and ♀ recognizable whole year by white specula, bordered anteriorly with chestnut feathers, which are clearest during flight. Bill of ♂ uniform grey, in ♀ and juvenile marked with orange and dark grey spots. Legs and feet orange yellow with black webbing. Voice: ♂ deep 'kack kack' and high-pitched 'quer-r-r', ♀ like female Mallard. Fairly common nester in some pond and lake regions with thick vegetation, in others almost completely absent. Breeding season: IV—VI. Nest like Mallard's hidden under dense vegetation. Eggs: 8—12, usually creamy yellow, incubated about 26 days. Diet: like Mallard's. Migrant, a small proportion partial migrants. Migration: IV and IX—XI.

3 **PINTAIL** *(Anas acuta)* Size of Mallard, but slimmer, with tapering tail, thinner neck and narrow bill. ♂ in nuptial plumage has white breast and distinctively marked head. ♀ tan, speckled, with grey bill. Winter plumage of ♂ and juvenile similar to ♀. Downy plumage closely resembles Mallard's. Specula bronze green in ♂, poorly developed in ♀. Identifiable in flight by light posterior border of wings. Legs grey. Voice: courting ♂ utters muffled 'quuck quuck', ♀ comparatively silent. Infrequent nester in pond and lake regions. Breeding season: IV—V. Nest like Mallard's, made of dry vegetation, on ground, in grass or under bush. Eggs: usually 7—11, elongate, yellowish, incubated 22—23 days by ♀. Young follow female about 7 weeks. Diet: mainly vegetarian. Migrant. Migration: IX—XI and III—IV.

1

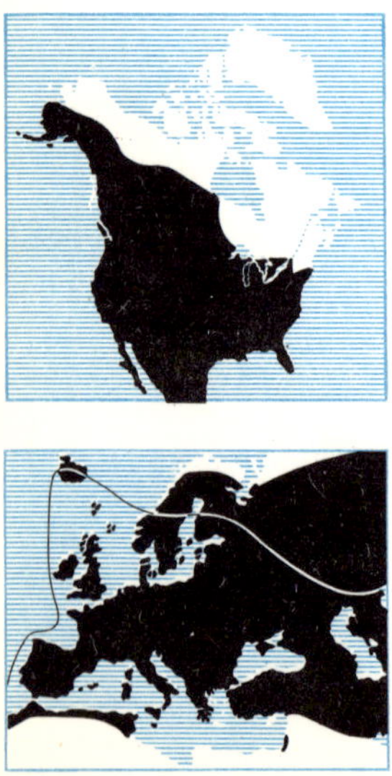

2

3

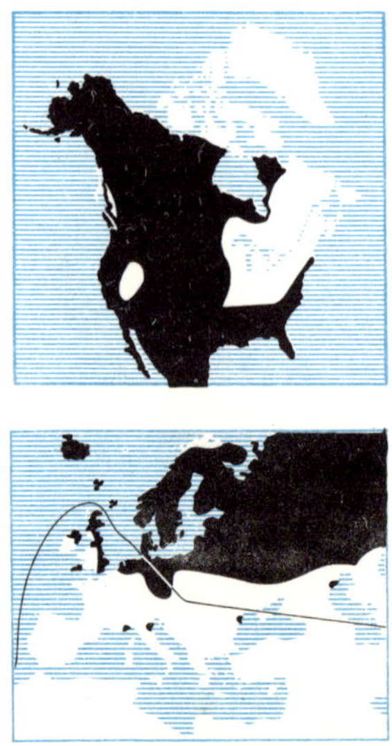

1 ♀
1
1 ♂ S
1 D
2 ♀
2
2 ♂ S
3
3 ♀
3 ♂ S

1 **SHOVELER** *(Anas clypeata)* Almost as large as Mallard, easily recognized by long, wide bill. ♂ in nuptial plumage has distinctive chestnut sides. Winter plumage, ♀ and juvenile similar to female Mallard. Legs orange red. Downy plumage resembles Mallard's. Voice: ♂ low, guttural 'took took', ♀ quacks like Mallard. Courting birds usually nod vigorously. Widespread, but not very common nester on fishponds and lakes encroached on by land. Breeding season: IV—V. Nest hidden in clump of sedge, shallow water, waterside vegetation or meadow, padded with greyish brown down. Eggs: 7—12, greenish grey, incubated about 25 days by ♀. Young follow female about 7 weeks. Food gathered mainly from water, strained through large, chattering bill. Diet: mostly small aquatic animals and seeds. Migrant. Migration: IX—XI and III—IV.

2 **TEAL - GREEN-WINGED TEAL** *(Anas crecca)* Smallest dabbling-duck. Recognizable even at long distances by narrow white shoulder stripe and yellow patch on underside of tail (♂). ♀ brown and finely speckled. Brilliant green specula, black-grey bill, dark brown eyes, dark grey legs. Flies easily and fast, with whirring wings, also very agile on ground. Downy plumage like Mallard's. Voice: ♂ 'krit', ♀ 'quackquack'. Widespread, but not common nester in inland regions with ponds and swamps, but one of most abundant ducks in summer and autumn owing to immigration. Winters mainly in western and southern Europe. Breeding season: IV—V. Nests in dense waterside vegetation. Eggs: 8—10, wide oval, yellowish or greenish grey, incubated 21—23 days by ♀. Young follow female 6—7 weeks. Diet: creatures living in mud, parts of plants, seeds. Partial migrant. Migration: IX—XI and III—IV.

3 **GARGANEY** *(Anas querquedula)* About size of Teal. ♂ in nuptial plumage distinguishable by light brown neck, wide, pure white eye stripe and trailing dorsal plumes. Winter plumage: ♀ and juvenile lighter in general than Teal, with bolder eye stripe. Bill and legs grey, eyes brown. Downy plumage like Mallard's. In flight more variegated than Teal, with blue-green scapulars and light green, white-edged specula. Voice: spring note of ♂ low, crackling sound, ♀ 'knāk'. Breeding season: IV—VI. Nest like Teal's, padded with dark brown down, with long, lighter tips. Eggs: 8—11, elongate oval, creamy yellow, incubated 21—23 days. Young follow adults about 6 weeks. Diet: parts of plants, small aquatic creatures. Migrant, winters largely in Africa. Migration: VII—IX and III—IV.

1

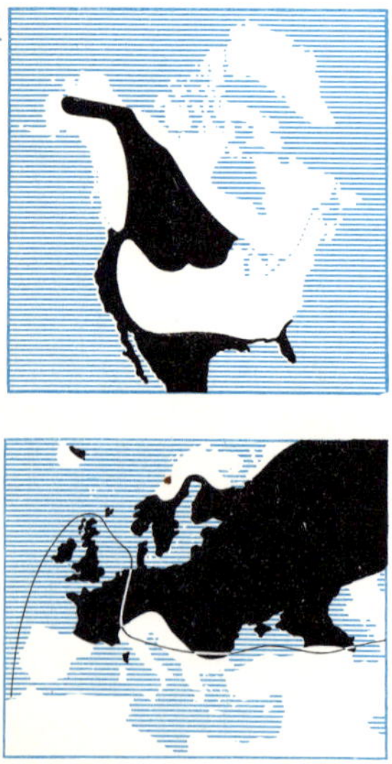

2

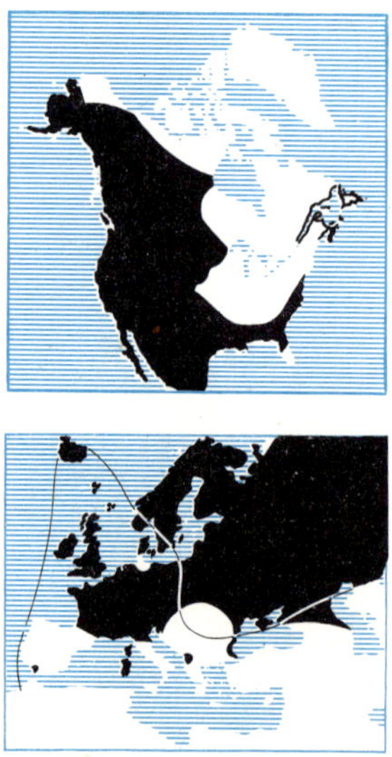

3

1
1 ♀
1 ♂ S
2
2 ♀
2 ♂ S
3
3 ♀
3 ♂ S

1 **WIGEON - EUROPEAN WIDGEON** *(Anas penelope)* Moderately large dabbling-duck with rather short, bluish green bill and round head (bulging forehead). ♂ in nuptial plumage has reddish brown head, cream crown stripe and claret breast. In winter, like ♀ and juvenile, reddish brown to cinnamon, with white scapulars. Legs grey, eyes brown. In flight shows white field on upper surface of wings (in ♀ grey). Downy plumage like Mallard's. Voice: ♂ whistling 'whee-ōō', ♀ purring growl, 'trrr'. Scandinavian species, nests in waterside marshes; during migration appears on ponds and in shallow bays. Breeding season: V—VII. Nest hidden in waterside vegetation, padded with ash-grey down. Eggs: 7—11, creamy yellow. Diet: almost entirely vegetarian. Migrant, winters largely in Africa and southern Asia. Migration: IX—XI and III—V.

2 **RED-CRESTED POCHARD** *(Netta rufina)* Robust diving-duck about size of Mallard. ♂ in nuptial plumage has distinctive crimson bill and golden brown crest. ♀ greyish brown, with light grey, clearly demarcated cheeks. Legs pink to scarlet, eyes reddish brown (♀) or scarlet (♂). In flight shows white posterior half of wings and ♂ broad black belly stripe. Downy plumage sepia brown on back, pale yellow on underside. Dives little. Voice: ♂ calls 'bāt'. Sporadic nester, in small communities, by warm, shallow lakes and ponds. Breeding season: IV—VI. Nest hidden in dense vegetation near water, lined with dry blades, leaves and light brown down. Eggs: 6—12, greyish yellow, incubated about 26 days by ♀. Young follow ♀ about 2 months. Diet: almost entirely vegetarian. Migrant. Migration: IX—XI and III—IV.

3 **GOLDENEYE - COMMON GOLDENEYE** *(Bucephala clangula)* Robust diving-duck with angular head and short bill. ♂ in nuptial plumage glossy black and white, with large oval spot on sides of black head. Winter plumage has narrow white neck band (absent in juvenile). Downy young vivid black and white. Legs mainly orange, eyes light yellow. Whirring, melodious flight. Voice: courting ♂ calls 'quee-reek', ♀ 'kurr kurr'. Scandinavian species, nests only here and there in central Europe, but regular winter bird on large rivers, dams and ponds. Breeding season: IV—V. ♂ courts ♀ with neck held stiffly erect. Nests in hollow tree up to 20 m above ground, also in nesting boxes. Eggs: 6—11, rounded oval, bluish green, packed in pale grey down, incubated 28—30 days. Diet: mainly animal (insect larvae, molluscs). Migrant. Migration: IX—XI and III—IV.

1
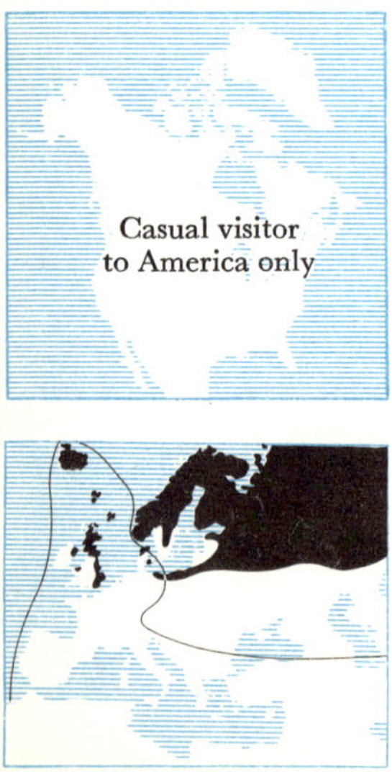

2

3
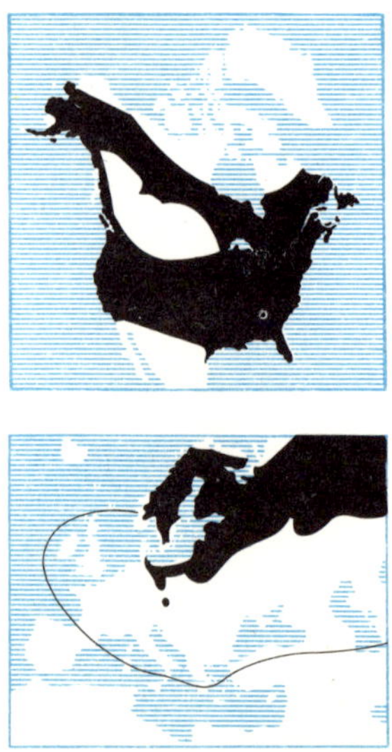

1
1 ♀
1 ♂ S
2
2 ♀
2 ♂ S
3
3 ♀
3 ♂ S
3 D

1 **TUFTED DUCK** *(Aythya fuligula)* Smaller than Mallard. Typical diving-duck, short tail sits almost on water. Dives often. ♂ in nuptial plumage deep black, with gleaming white, sharply defined sides, thin, drooping crest on back of head and blue-grey bill. ♀ dark brown, with darker sides, off-white belly and rudimentary crest. ♂ in winter plumage resembles ♀. White posterior wing border visible during flight. Eyes yellow, legs grey and black. Downy and juvenile plumage deep brown to black. Voice: harsh 'kur-r-r' when flying, courting note soft, whistling 'tschihihihi'. Common summer bird on fishponds and lakes and on some city ponds. Frequent winter bird on rivers. Breeding season: V—VII. Often nests in Black-headed Gull colonies. Eggs: 8—12, elongate, grey-green, incubated about 25 days. Diet: chiefly animal. Migrant. Migration: IV and IX—X.

2 **SCAUP - GREATER SCAUP** *(Aythya marila)* Always distinguishable from Tufted Duck by thicker neck and rounded nape. ♂ in nuptial plumage has light grey back, ♀ usually identifiable by broad white ring at base of bill. Eyes brown to light amber. Downy and juvenile plumage like Tufted Duck's. White-bordered wings in flight. Voice: soft whistling and cooing (♂) and raucous warning 'karr karr' (♀) at nesting site. Nests by lakes and in bogs. In winter frequents coasts, bays, river mouths and (mainly in severe winters) lakes and rivers. Breeding season: VI—VII. Nests on ground near water. Eggs: 6—10, elongate oval, greenish brown, packed in dark down, incubated about 25 days. Diet: aquatic animals, some vegetable matter. Migrant. Migration: IX—XI and III—IV.

3 **POCHARD** *(Aythya ferina)* Smaller than Mallard. Typical diving-duck, short tail lies on water. ♂ in nuptial plumage has russet head and neck and silvery grey back. ♀ mainly light brown, with grey-brown back. Has flatter forehead than other diving-ducks. Eyes orange red, beak dark grey and blue (anterior half), feet dark grey. Downy plumage yellowish brown. Voice (heard usually only at mating time): soft 'wibwibwib', descending 'hihihi' (♂) and rattling 'kur-r-r' (♀). Lively courtship by ♂♂, with outstretched neck, while ♀♀ swim about. Frequent summer bird on fishponds and lakes, in winter seen on reservoirs and ice-free rivers. Breeding season: IV—VI. Nest hidden in waterside vegetation. Eggs: 6—9, often broad oval, greenish or yellowish grey, packed in down, incubated about 25 days. Diet: aquatic animals and plants. Migrant. Migration: IX—XI and III—IV.

1

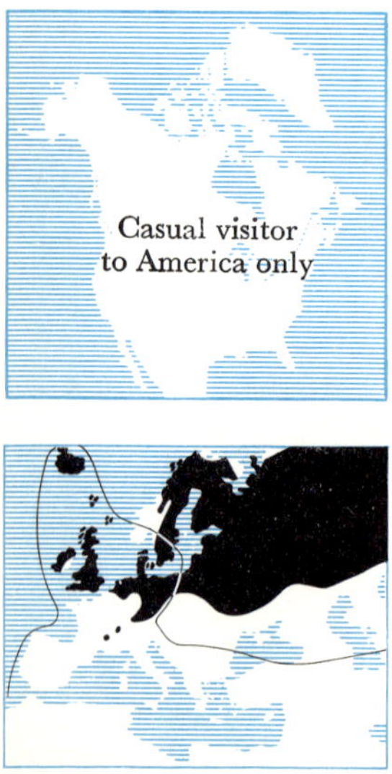

2

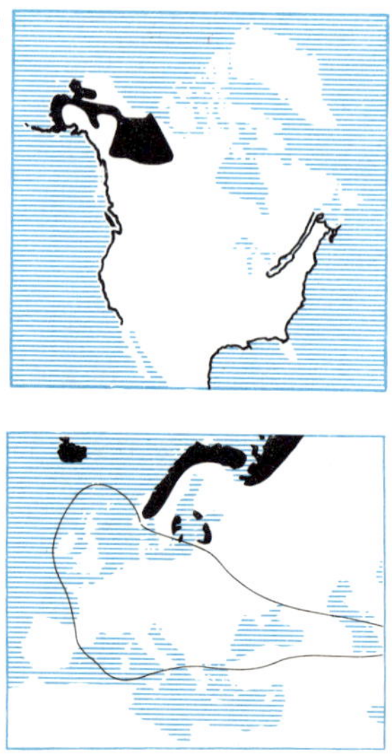

3

1
1 ♀
1 ♂ S
1 D
2
2 ♀
2 ♂ S
3
3 ♀
3 ♂ S
3 D

1 **FERRUGINOUS DUCK** *(Aythya nyroca)* Small pochard with mainly chestnut plumage. Distinguished from other diving-ducks by white specula, pure white under tail-coverts and (in flight) broad, gleaming white posterior wing border. Eyes white, bill mainly dark grey, legs grey. ♂ and ♀ alike, in summer and winter. Downy plumage sepia brown, with yellow underside. Voice: audible only at mating time and at close range. Not very common summer bird on marshy ponds and lakes. Breeding season: V—VI. Nest hidden in waterside vegetation. Eggs: 7—11, blunt oval, yellowish brown, packed in dark down, incubated 25 days. Diet: mainly parts of aquatic plants, also molluscs, crustaceans and insect larvae. Migrant. Migration: X—XI and III—IV.

2 **LONG-TAILED DUCK - OLDSQUAW** *(Clangula hyemalis)* Smallest diving-duck, with small, round head, short bill and wedge-shaped tail terminating in 'stiletto' feathers (♂). Dives often and for long periods. Downy plumage blackish brown. Voice: ♂ has sonorous, rising call of 3—4 syllables, 'ow-ow-owdlow'. Flies rapidly, recognizable by plain dark wings, spotted head, white belly and (♂) black stripe down middle of back. Nests by fresh water in arctic tundra. Winters regularly on coasts of western and northern Europe (a few birds inland). Breeding season: VI—VII. Nest near water, usually concealed by vegetation. Eggs: 5—9, elongate oval, yellowish or greenish grey, packed in down, incubated about 25 days. Diet: mainly molluscs and crustaceans. Migrant. Migration: IX—XI and III—IV.

3 **EIDER - COMMON EIDER** *(Somateria mollissima)* Larger than Mallard. Has distinctive long, triangular head and bill, low forehead. Unlike other species, ♂ in nuptial plumage has black belly and white back, in winter plumage ± black. ♀ brown, with dense black striping. Downy plumage dark brown. Recognizable in flight by robust form and short, thick neck. Frequent diver. Voice: call of ♂ resounding 'ah-ōō' or 'cōō-rōō-uh', ♀ utters cackling and crackling sounds. Nests only on coast, lives in off-shore waters, occasionally winters inland on fresh water. Breeding season: IV—VI. Courtship usually in groups of several ♂♂ and one ♀. Nest among boulders or in littoral vegetation, with a thin framework, but thickly padded with down. Eggs: 4—6, fairly large, greenish grey, incubated about 25 days. Young follow adults 9—10 weeks. Diet: chiefly molluscs. Resident bird or migrant. Migration: X—XI and IV.

1

2

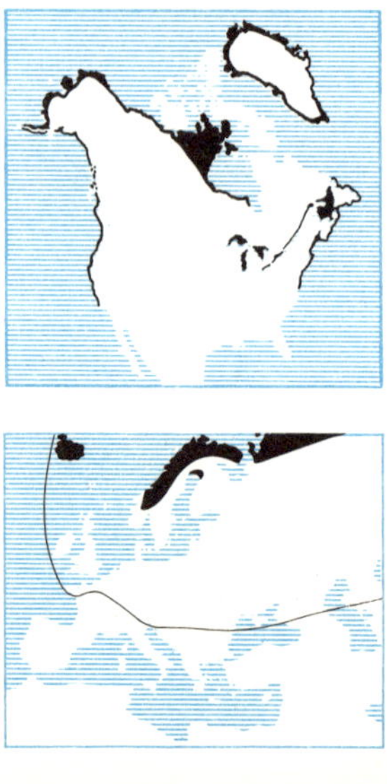

3

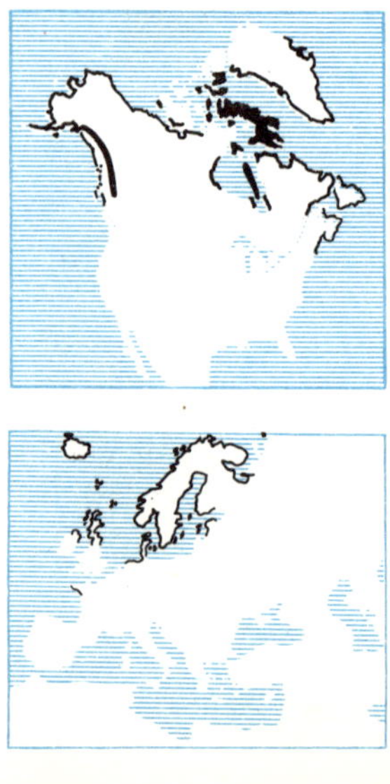

1
1 ♀
1 ♂ S
2
2 ♀
2 ♂ S
3
3 ♀
3 D
3 ♂ S

1 **VELVET SCOTER - WHITE-WINGED SCOTER** *(Melanitta fusca)* Ungainly black duck about size of Mallard. ♂ in nuptial plumage deep black, with small white spot behind eyes and white specula. ♀ and juvenile blackish brown with two white spots on either side of head. Legs red. Always distinguishable from Common Scoter by white specula. Identifiable in flight by conspicuous white bands on secondaries. Downy plumage dark brown. Dives often, with half-spread wings. Voice: courting ♂♂ utter rather loud 'who-er', ♀ growls; in winter usually silent. Nests by Scandinavian lakes and Baltic and on off-shore islands. Winters mainly off shore, but a few birds regularly winter on inland waters. Breeding season: mostly VI. Eggs: 7—10, light brown, incubated about 28 days. Diet: molluscs. Migrant. Migration: X—XI and III—IV.

2 **COMMON SCOTER** *(Melanitta nigra)* Brownish duck, smaller than Mallard. ♂ in nuptial plumage all deep black, ♀ and juvenile blackish brown, with grey-sided head and black crown. Legs black. Closely resembles Velvet Scoter, but lacks white markings (also in flight). Downy plumage smoky black and brown. Dives frequently with bill half open. Voice: courting ♂ utters fluty 'coorlee', ♀ raucous 'kur-r-r'. Nests beside Scandinavian inland waters, preferably on small islands. Otherwise found mainly on sea, near shore; sometimes winters inland. Breeding season: mostly VI. Nest hidden in waterside vegetation; small hollow padded with dark brown down. Eggs: 6—9, light reddish brown, incubated 28—30 days by ♀. Diet: mainly molluscs, fewer crustaceans and worms. Migrant. Migration: IX—XII and III—IV.

3 **SHELDUCK** *(Tadorna tadorna)* Large goose-shaped duck. Vivid white, black and russet markings and red beak. ♂ in nuptial plumage has prominent knob on beak (absent in ♀). Legs flesh pink. Juvenile has greyish white head and lacks russet breast band. Downy plumage white, with vivid blackish brown markings. Does not dive, dabbles for food. Light on feet. Voice: calls 'ak-ak-ak' and raucous 'kor-r', but usually fairly silent. Inhabits coast, especially mud- and sand-banks in off-shore waters, seldom seen inland. Breeding season: IV—VI. Nests mostly in holes in ground, often in rabbit burrows. Eggs: 8—12, rounded oval, creamy white, packed in thick greyish white down, incubated 28—30 days by ♀. Young cared for and led 6—9 weeks by both parents. Diet: molluscs, crustaceans. Nomadic bird and migrant. Migration: VIII—XI and III—IV.

1

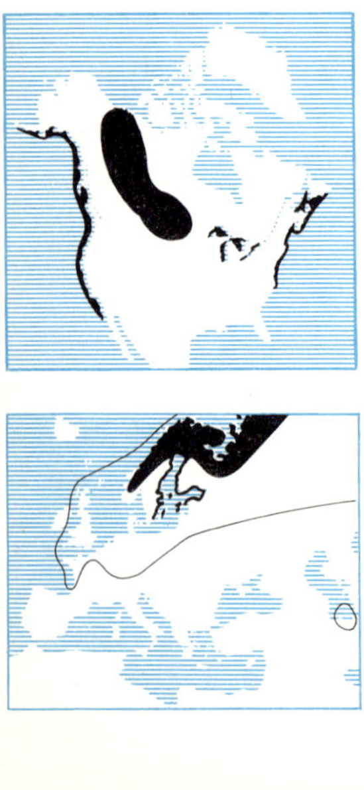

2

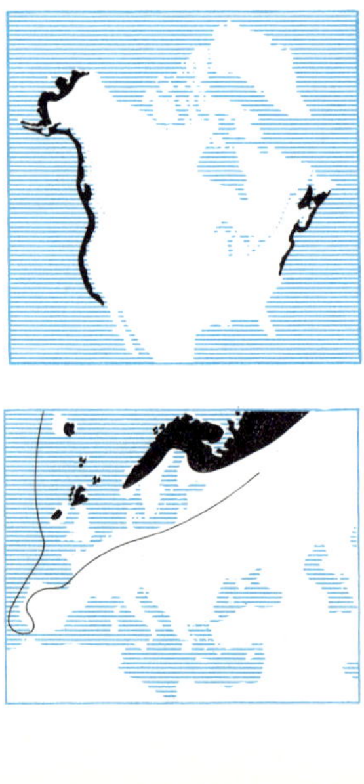

3

1 ♀
1
1 ♂ S
2
2 ♀
2 ♂ S
3
3 juv
3 ♂
3 D

1 **GOOSANDER - COMMON MERGANSER** *(Mergus merganser)* Larger and slimmer than Mallard, with narrow red bill. Swims deep in water, dives frequently. ♂ in nuptial plumage has white breast and sides (usually with salmon pink tinge), head and top of neck bottle green. ♀ and juvenile mainly ash grey, with reddish brown head and neck, characteristic crest and sharply circumscribed white chin spot (cf. Red-breasted Merganser). Downy plumage brown on head, black on back, white on underside. Voice: in winter mostly silent, at mating time ♂ utters grating 'kerr kerr'. Scandinavian species, regularly appears on ice-free rivers in winter. Breeding season: III—V. Nest in hollow tree up to 15 m above ground. Eggs: 8—12, creamy yellow, packed in white down, incubated 32—35 days. Diet: almost entirely fish. Partial migrant. Migration, depending on ice, in X—XI and III.

2 **RED-BREASTED MERGANSER** *(Mergus serrator)* Almost as large as Mallard, with thin neck, long red bill and ragged crest. ♂ in nuptial plumage distinguished by rust brown breast band. ♀ and juvenile closely resemble Goosander (differences: size, shape of bill and crest, blurred outlines of white chin spot). Legs red, eyes reddish yellow. Downy plumage like Goosander's. Voice: courting ♂♂ call 'da-ah', ♀♀ hoarse 'rokrokrok'. Scandinavian species, nests mainly on coast and in river mouths, also winters on inland waters. Breeding season: IV—VI. Nest on ground, near water, hidden in dense vegetation. Eggs: 6—12, olive green, packed in light brown down. Young fledged at 8—9 weeks. Diet: chiefly fish, also crustaceans, worms, insects. Partial migrant. Migration: IX—X and IV—V.

3 **SMEW** *(Mergus albellus)* Small duck with short bill. ♂ in nuptial plumage pure white except for black stripe on head and back and dark primaries. ♀ and juvenile grey, with reddish brown crown and white cheeks. In flight mostly white. Flies fast and silently, with quick wing strokes. Beak dark grey, legs lead grey, eyes brown (♀) or grey (♂). Downy plumage vivid black and white, closely resembles Goldeneye. Voice: ♂ utters rattling and growling mating calls. Arctic species, inhabits inland water, comparatively rare winter bird on ice-free rivers, lakes and ponds. Breeding season: V—VI. Nest in hollow tree, lined only with greyish white down. Eggs: 6—9, creamy yellow, incubated about 30 days. Diet: mainly small fish, also insects, crustaceans, worms. Migrant. Migration: IX—XII and III—IV.

1

2

3

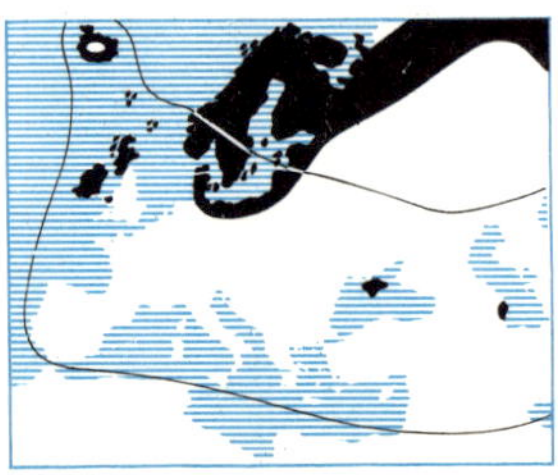

1
1 ♀
1 D
1 ♂ S
2
2 ♀
2 ♂ S
3
3 ♀
3 ♂ S

1 **GREYLAG GOOSE** *(Anser anser)* Size of domestic goose, light brownish grey. Distinguished from other geese by large orange yellow to pink bill with white nail and by pale grey anterior border of wings. ♂ = ♀. Downy plumage olive green on back, yellow on underside. Flies with slow, powerful wing strokes, if in numbers in echelon formation. Gregarious bird. Voice: nasal cackle 'aahng-ung-ung' and honking advertisement call like Domestic Goose. Rare summer bird on a few quiet lakes and ponds. Breeding season: III—IV. Nest on pile of crushed reeds and sedge, hidden in rushes, padded with light grey down. Eggs: 4—8, large, dull white, incubated 28—29 days by ♀, guarded by ♂. Diet: vegetarian, mainly grasses (cereal shoots, etc.). Partial migrant. Migration: IX—X, and II—III.

2 **WHITE-FRONTED GOOSE** *(Anser albifrons)* Slightly smaller than Greylag Goose, with uniformly coloured bill. Adults have white 'blaze' and black abdominal bars; closely resemble Lesser White-fronted Goose. Juveniles have no 'blaze'; distinguishable from Lesser White-fronted Goose by longer bill and greater size. Anterior part of wings grey in flight (cf. Greylag Goose). Has flesh pink bill, orange red legs, dark brown eyes and dark orbital ring. ♂ = ♀. Downy plumage sepia brown on back, golden yellow to brown on underside. Voice: high, non-nasal call of two syllables 'kowlyok'. Inhabits tundra of northern Russia, winters mainly in coastal areas. Breeding season: VI—VII. Eggs: 5—6, pale yellow. Habits and diet like Greylag Goose. Migrant. Migration: X—XII and III—IV.

3 **LESSER WHITE-FRONTED GOOSE** *(Anser erythropus)* Smaller than White-fronted Goose, but in all plumages so similar that distinguishable only under favourable conditions. Adults usually have larger 'blaze', extending between their eyes. Head more rounded, bill shorter, orbital ring bright yellow (also juveniles), abdominal bars fainter and wing tips longer (extend just beyond tip of tail when goose is standing). Voice: call higher and faster than White-fronted Goose's, usually 'kyu-yu-yu' or 'kyu-yut'. Nests in arctic tundra. A few birds winter in central Europe, but the majority in eastern Europe, with other geese. Habits like White-fronted Goose's. Migration: VIII—X and III—IV.

1

2

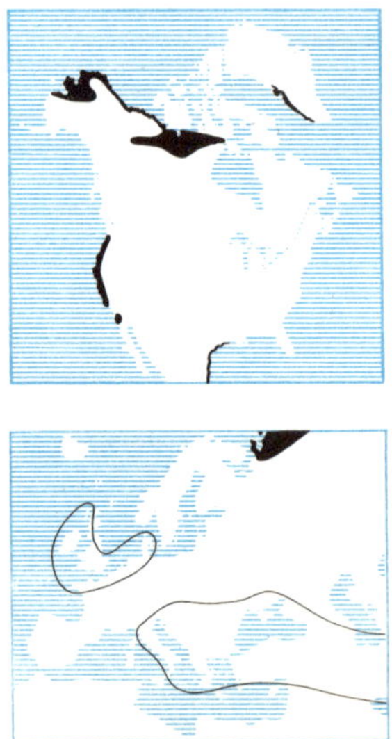

3

1 D
1
2 juv
2
2 ad
3
3 juv
3 ad

1 **BEAN GOOSE** *(Anser fabalis)* Rather smaller than Domestic Goose. In general darker than Greylag Goose, especially head and neck. Anterior half of wings dark (seen in flight). Bill half black (at base), half yellow, with black nail. Legs orange yellow, eyes brown. ♂ = ♀. Downy plumage resembles Greylag Goose's. Voice: nasal 'kajak, kajaiak' and 'ung' calls when flying. Nests in Scandinavian wooded tundra, comes in large flocks to central and western Europe for winter. Breeding season: V—VI. Nest depression in dry ground, thinly lined with grass and moss. Eggs: 4—6, dull white, incubated 27—29 days. Young cared for by both parents, fledged at about 2 months. Diet: in winter quarters same as for Greylag Goose. Migration IX—X and II—III.

2 **BRENT GOOSE - BRANT GOOSE** *(Branta bernicla)* Small, very dark goose about size of Mallard, with short black bill and black legs. Distinguished from Barnacle Goose by completely black head and narrow white mark on either side of neck. Recognizable in flight by black head, neck and breast. Downy plumage mainly grey, brownish grey down middle of back. Voice: mostly deep, throaty 'rrouk' and 'rruk' calls. Nests in arctic tundra, winters regularly in western and northern Europe, when especially numerous in shallow off-shore waters in North Sea; occasionally occurs inland (grey- and light-bellied varieties). Breeding season: VI—VII. Nest made of moss, lichen and grass on dry ground near water, padded with down. Eggs: 3—5, pale grey, incubated about 25 days. Diet: grasses and lichens, in winter quarters mainly Zostera weed. Migration: IX—XI and III—IV.

3 **BARNACLE GOOSE** *(Branta leucopsis)* Moderately large goose with black and white markings, short black tail and black legs. Clearly distinguishable from Brent Goose by white face, banded grey back and (in flight) by white belly and black under surface of wings. Downy plumage light variegated brown. Voice: barking, rapidly repeated 'gnuk' calls (in flight), otherwise muffled 'óg, ark'. Nests in Scandinavian arctic tundra, mostly on steep rocks near coast. Regular winter bird in southern parts of North Sea, bird of passage in Baltic region. Seldom seen inland. Breeding season: V—VI. Nest shallow depression lined only with down and ringed by characteristic mound formed of sitting bird's excreta. Eggs: 4—6, greyish white, incubated about 25 days. Care of young about 7 weeks. Migrant. Migration: X—XII and III—IV.

1

2

3

1
2
2 light-bellied variant
2 grey-bellied variant
3
3

1 **MUTE SWAN** *(Cygnus olor)* Long, usually arched neck, orange beak with knob, grey-black legs, brown eyes. Juvenile white and greyish brown, with grey to flesh-pink beak. Downy plumage greyish brown. When displaying or threatening, retracts head and spreads wings like sails. Flies with slow, powerful wing strokes, making rhythmical sound audible for long distances. Voice: mostly silent. Distribution mainly concerns semi-domesticated birds frequenting lakes, ornamental ponds, fishponds, shallow bays and, in winter, open rivers. Breeding season: IV—V. Nest mostly large structure made of brushwood, sedge and rushes. Eggs: 5—8, large, dull brownish yellow, incubated 35—38 days by ♀. Diet: aquatic plants. Partial migrant, often nomadic out of breeding season.

2 **WHOOPER SWAN** *(Cygnus cygnus)* Distinguished from Mute Swan by straight neck and lemon yellow, black-tipped beak with no knob. Yellow part of beak tapers off into black tip. Flies quietly and does not adopt special menacing attitude. Legs black, eyes dark brown. ♂ = ♀. Juvenile and downy plumage like Mute Swan's. Voice: regular whooping call of 2—3 syllables, 'whōōp, whōōkoock', often rising to trumpeting call during flight. Scandinavian summer bird, nests on lakes in tundra. Regular winter bird of North Sea and Baltic coasts, also appears in small numbers on ice-free water inland. Breeding season: IV—VI. Nest like Mute Swan's. Eggs: 4—7, elongate oval, cream, incubated at least 31 days by ♀. Diet: like Mute Swan's. Migrant. Migration: X—XII and II—III.

3 **BEWICK'S SWAN - WHISTLING SWAN** *(Cygnus bewickii)* Closely resembles Whooper Swan, smaller size noticeable only on close comparison. Yellow part of beak does not taper forward as in Whooper Swan and neck distinctly shorter and thicker. Otherwise, all plumages and behaviour very similar. ♂ = ♀. Voice: loud call of one syllable 'hoo' and 'howk', strung together when the bird is excited ('hukkukkukkuk'); also calls while flying. Nests in tundras of northern Russia, regularly winters in small flocks on North Sea coast, bird of passage in Baltic. Also appears in small numbers on rivers. Breeding season: V—VI. Nest pile of stalks and moss on marshy ground. Eggs: 3—5, cream, incubated about 30 days; young learn to fly at 1½ months. Winter bird. Migration: X—XII and III—IV.

1

2

3

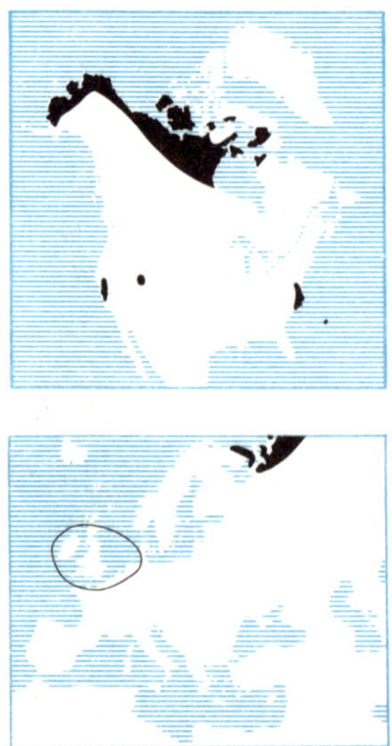

1
1 juv
1 D
1 ad
2
2 juv
2 ad
3 ad

Order: Birds of Prey - Falconiformes **Family: Vultures, Eagles, Buzzards, etc.** - Accipitridae

1 **GRIFFON VULTURE** *(Gyps fulvus)* Imposing vulture with long, wide, round-ended wings and spread primaries. Wing span about 2.2 m. Tail short and square-ended. Head and neck covered with white down. ♂ = ♀. Juvenile has brown ruff. Voice: soft chattering, crowing and hissing. Circles hours at great heights, hardly moving wings, looking for food. Gregarious, mainly inhabits mountains in Mediterranean countries, wanders far afield in search of food. Breeding season: II—IV. Nest: in crevice in rock face, sometimes on tree. Egg: one, white, incubated 51—52 days by ♀. Care of nestling about 3 months. Diet: large mammal carcases. Resident and nomadic bird, occurs in central Europe only as accidental bird.

2 **BLACK VULTURE** *(Aegypius monachus)* Slightly larger than Griffon Vulture (wing span about 2.5 m), with larger head, stronger beak, naked bluish neck, blackish brown plumage and longer, wedge-shaped tail (seen in flight). ♂ = ♀. Juvenile darker than adult. Voice: like Griffon Vulture's. Inhabits mountains and foothills with a few old trees and, in warm countries, steppes. Habits similar to Griffon Vulture's, but less gregarious and more arboreal. Seen in central and western Europe as accidental bird. Breeding season: II—IV. Nest: large structure of sticks and brushwood, usually on tall tree, seldom on rocks. Egg: one, dull white with a few red-brown spots, incubated over 50 days by both parents. Care of nestling about 3 months. Diet: carrion (large mammals). Resident and nomadic bird.

3 **WHITE-TAILED EAGLE** *(Haliaeëtus albicilla)* Very large eagle. Wing span up to 2.4 m. In flight, wide wings like boards, with spread primaries. Adults have bright yellow, prominent hooked beak and short, white, wedge-shaped tail, juveniles dark beak and ± dark tail. Legs featherless, lemon yellow. Downy plumage brownish grey. Voice: at mating time utters jubilant 'krikrikri-krijagjagjagljow', lifting head. Found in coastal regions, near large rivers with good fish supply and by lakes. Breeding season: II—IV. Nest: huge eyrie made of sticks, twigs and brushwood in top of tall tree. Eggs: two, chalky white, incubated about 40 days, mainly by ♀. Care of nestlings about 10 weeks. Diet: fish, supplemented by birds, mammals and dead game. Partial migrant. Migration: X—XI and II—III.

1

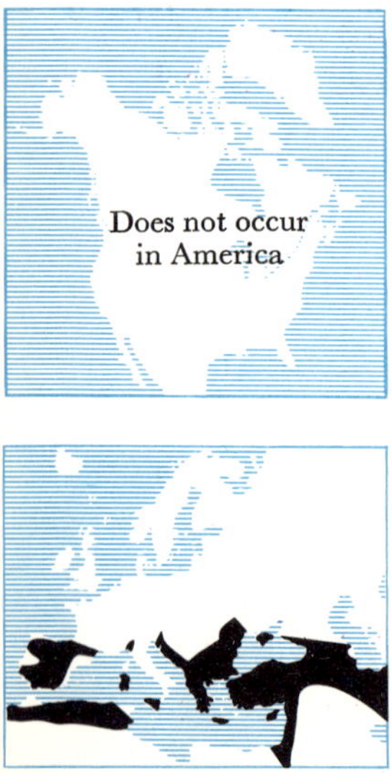

2

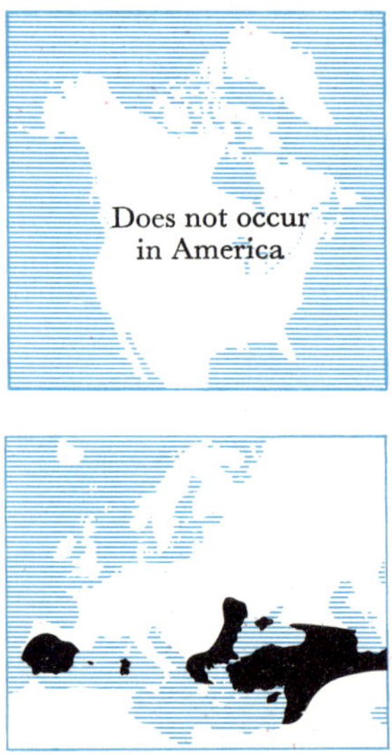

3

1
1 juv
1 ad
2 ad
2
3 juv
3 ad
3

Order: Birds of Prey - Falconiformes **Family: Vultures, Eagles, Buzzards, etc. -** Accipitridae

1 **GOLDEN EAGLE** *(Aquila chrysaëtos)* Large eagle. Wing span 195—220 cm. Distinguished from White-tailed Eagle by longer, rounded tail and dark, thinner beak. Legs feathered down to toes. Juvenile has white, black-banded tail and large white spot on under surface of wings. Adults generally one colour, with dark tail and golden, lanceolate feathers on crown and nape. Downy plumage white. Voice: hoarse 'twee-o' and barking 'kja' calls. Rare summer bird in high mountains, also occurs in wooded plains out of breeding season. Breeding season: III—IV. Nest: mostly on ledge of rock. Eggs: two, dull white with grey-brown spots, incubated 40—45 days by both parents. Care of nestlings up to 80 days. Diet: largish mammals and birds. Resident and nomadic bird.

2 **IMPERIAL EAGLE** *(Aquila heliaca)* Large, rather ungainly-looking eagle. Wing span 170 to 200 cm. Adults distinguished from Golden Eagle by white scapulars and generally darker plumage (except for light yellow crown and nape). Juveniles have honey-coloured or blackish brown spots. Compared with Golden Eagle in flight, cross-banded tail looks shorter and wings terminate in 7 spread primaries (in Golden Eagle 5). Downy plumage white. Voice: barking calls during courtship. Rare summer bird in open flat and hilly country, wooded steppes and valleys. Breeding season: IV. Nest: large eyrie made of broken branches and brushwood. Eggs: 2—3, greyish white with faint brown spots, incubated about 43 days, chiefly by ♀. Care of nestlings about 2 months. Diet: mainly small mammals (ground squirrels), occasionally carrion. Nomadic bird and migrant. Migration: X—XI and III.

3 **SPOTTED EAGLE** *(Aquila clanga)* Medium large, dark brown eagle with 7 visible, spread primaries and rather short, slightly rounded tail. Wing span 160—180 cm. In juvenile, dense white spots on upper surface of wings look like two white bands at a distance. Closely resembles Lesser Spotted Eagle, but distinguished by brighter white spot on rump. Downy plumage white. Voice: loud 'kyakyakyak' calls like yapping and protracted 'eaf eaf'. Rare summer bird in woods in open country with stretches of water. Accidental bird in central and western Europe. Breeding season: IV—V. Nest: fairly large, made of branches and brushwood in top of old deciduous tree. Eggs: 1—2, greyish white, sparsely spotted. Incubation and hatching period as for Imperial Eagle. Diet: small rodents, reptiles, frogs, birds, occasionally carrion. Migrant and nomadic bird. Migration: X—XI and IV.

1

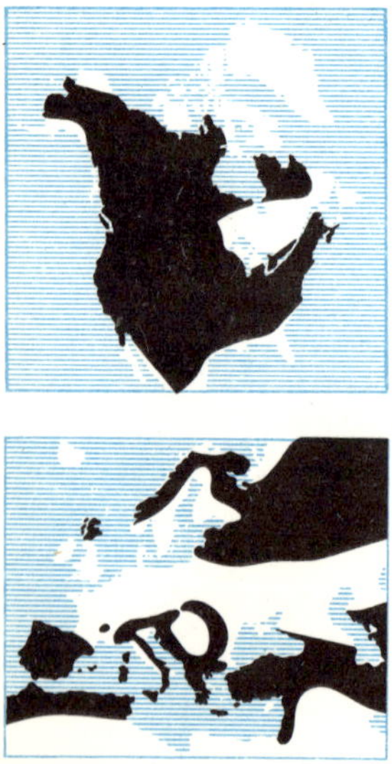

2

3

1 juv
1
1 ad
2
2 ad
2 juv
3 ad
3
3 juv

Order: Birds of Prey - Falconiformes **Family: Vultures, Eagles, Buzzards, etc. -** Accipitridae

1 **LESSER SPOTTED EAGLE** *(Aquila pomarina)* Moderately large eagle. Wing span 145 to 160 cm. Closely resembles Spotted Eagle, but smaller and commoner. Adults have lighter colouring and dark rump. Juveniles have light brown nuchal spot and are more finely speckled, their wings appearing to have several indistinct light bands in flight. Downy plumage white. Voice: various whistling calls. Rare summer bird in large forested regions with old trees and mixed woods with grassy clearings. Has undulating courtship flight with long swoops over the nest. Breeding season: IV—V. Nest: in top of tall tree. Eggs: 1—2, greyish white with dark brown spots, incubated about 43 days. Care of nestlings 7—8 weeks. Diet: frogs, snakes, lizards, mice (often caught on foot). Migrant. Migration: X—XI and III—IV.

2 **BOOTED EAGLE** *(Hieraaëtus pennatus)* Almost as large as buzzard. Wing span 110—120 cm. Might be mistaken for Honey Buzzard in flight, but has typical eagle features: prominent head, long wings with 6 spread primaries, and long, straight-edged tail. Two colour variants: light phase with dark wings, white underside and light brown tail (commoner), and with dark brown underside. Juvenile has rust-coloured belly. Voice: clear whistle, 'kee, keekeekeek'. Rare summer bird in deciduous and mixed woods in plains and mountains. Breeding season: IV—V. Nest: usually at top of deciduous tree. Eggs: 2, greenish white, sometimes speckled reddish brown, incubated about 1 month by ♀. Care of nestlings about $1\frac{1}{2}$ months. Diet: small land vertebrates and large insects. Migrant. Migration: IX—XI and IV.

3 **SHORT-TOED EAGLE** *(Circaëtus gallicus)* Larger than buzzard. Wing span 150—170 cm. Resembles Honey Buzzard in flight because of rather long wings and tail, but has much larger, round head. Underside almost entirely white, as in Osprey. Upper breast dark, tail marked with 3—4 indistinct dark bands. Juvenile like adult, but more streaked on underside. Beak bluish grey, legs light grey, eyes yellow to orange. Downy plumage white. Voice: frequent, loud 'mew-ok' or 'eeok-ok' (♂) and miaowing (♀). Inhabits warm regions with snakes, mountain valleys and ravines. Nests on trees. Breeding season: V. Nest like buzzard's, always with outer lining of green twigs. Egg: one, chalky white, incubated 45 days by both parents. Care of nestlings 70—80 days. Diet: almost nothing but snakes, sometimes detected by hovering. Migrant. Migration: IX—X and IV—V.

1

2

3

1 ad
1
1 juv
2 LV
2 DV
3 DV
3 LV

Order: Birds of Prey - Falconiformes **Family: Vultures, Eagles, Buzzards, etc. -** Accipitridae

1 **BUZZARD** *(Buteo buteo)* Larger than crow. Wing span 110—135 cm. Distinguished in flight by broad wings, almost closed primaries, round, scarcely projecting head and relatively short, fan-like tail with numerous narrow cross-bands. Beak thin, eyes brown to yellow, legs yellow and unfeathered. ♂ = ♀. Downy plumage white or grey. Voice: protracted 'peeiōō'. Inhabits forests adjoining fields. Commonest of large birds of prey. Breeding season: III—V. Swoops over nest at mating time. Nest: usually at top of tree. Eggs: 2—4, chalky white with grey and brown spots, incubated 28—31 days by both parents. Care of nestlings 6—8 weeks. Diet: mainly mice, hunts in open country, usually watches from look-out or flies slowly and observantly. Partial migrant, frequent winter bird. Migration: IX—XI and III—IV.

2 **ROUGH-LEGGED BUZZARD - ROUGH-LEGGED HAWK** *(Buteo lagopus)* Closely resembles Buzzard in shape and size, but distinguished by usually light underside of wings and tail, sharply outlined spots on carpal joint, wide black band at end of tail and dark patch on belly. Head white, with dark stripes, rump always white, legs feathered down to toes. Juvenile has wider, light border on dorsal feathers and light breast with wider streaks than adult. Habits and diet like Buzzard's, but hovers more when hunting. Nests in Scandinavian tundra, regular winter bird in western and central Europe. Breeding season: V—VI. Nest: usually on ground. Eggs: 3—4, like Buzzard's, incubated about 31 days by ♀. Care of nestlings 6—7 weeks. Migrant, irruptive bird in central Europe in some winters. Migration: X—XI and II—IV.

3 **LONG-LEGGED BUZZARD** *(Buteo rufinus)* Slightly larger than Buzzard. Striking whitish, streaked underside and russet under tail-coverts. Back rust brown, head usually lighter. Tail plain (Buzzard often similar colour, but has banded tail). Inhabits dry steppes and bare mountains, not woods. Occurs in central and western Europe as accidental bird. Nest: on ground, steep banks and rocks, occasionally trees, made of brushwood. Eggs: 2—4, chalky white, brown-speckled. Habits and diet like Buzzard's. Partial migrant. Migration: IX—XI and III—IV.

1

2
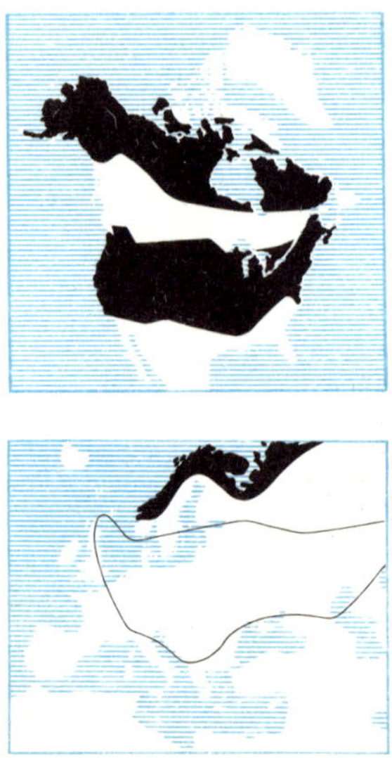

3

1 LV
1
1 DV
2
3
3
2

Order: Birds of Prey - Falconiformes **Family: Vultures, Eagles, Buzzards, etc. -** Accipitridae

1 **HONEY BUZZARD** *(Pernis apivorus)* Average buzzard size. Wing span 115—125 cm. Looks slimmer than Buzzard when flying, with narrower wings and longer tail. Small head, with narrow beak, tapering forwards like pigeon's. Has 3 distinct tail bands — a wide one at end and 2 narrow ones near base. Plumage very variable. Juveniles distinguished by yellow-white spots on back and frequently almost white head. Downy plumage white. Inhabits small deciduous woods and pine-woods with adjoining meadows and fields. Breeding season: V—VII. Nest like Buzzard's. Eggs: 2, rounded, with thick brown speckling, incubated 35—40 days by both parents. Diet: mainly wasps' larvae, scratched out with feet, also small vertebrates. Migrant. Migration: IV—V and VIII—X.

2 **SPARROW HAWK** *(Accipiter nisus)* Size of Collared Turtle Dove. Wing span 60—80 cm. ♀ ¼ larger than ♂. Distinguished in flight by short, broad, rounded wings (cf. pointed wings of similarly sized Kestrel) and long, straight-ended tail with 4 bands. Also characterized by rapid wing strokes, alternating with gliding. Juvenile distinguished from grey-backed adult by brown back. Legs yellow, eyes light yellow. Downy plumage white. Voice: penetrating 'kekkekkek' calls near eyrie. Common summer bird in conifer woods adjoining open country with hedges, groups of trees and gardens. Breeding season: V—VI. Nest: small, made of twigs, not too high on tall pine. Eggs: 4—6, rounded, with thick brown speckling, incubated 33 days by ♀. Care of nestlings 28 days. Diet: mainly small birds. Partial migrant. Migration: IX—X and III—IV.

3 **GOSHAWK** *(Accipiter gentilis)* About size of buzzard (♂ smaller). Wing span 100—120 cm. Resembles Sparrow Hawk in flight, but larger. Dense bands on underside always blackish. Has conspicuous white under tail-coverts. Juvenile has large dark streaks on underside, instead of wavy bands. Legs yellow, eyes orange red to yellow. Downy plumage white. Voice: shrill serial 'giggiggiggiggiakgiak' on eyrie. Attacks prey from ambush and chases it in agile flight. Common summer bird in large forest regions. Breeding season: IV—VI. At mating time, ♂ swoops up and down over eyrie. Nest made of broken branches and brushwood near top of tree, always lined with green twigs. Eggs: 3—4, rounded, greenish white, usually plain. Incubated 36 days by ♀. Care of nestlings 38 days. Diet: mostly largish birds, also mammals, especially squirrels. Resident bird, partial migrant in northern Europe only.

1

2

3
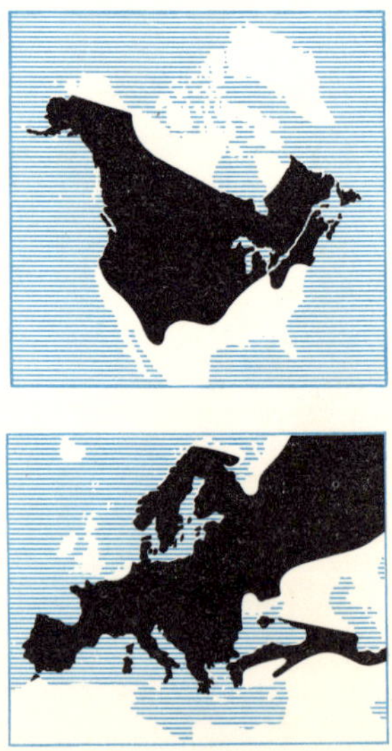

1 LV
1 DV
1 juv
1
2 ♀
2 ♂
2 juv
3 ♀
3
3 juv

1 **RED KITE** *(Milvus milvus)* Size of large buzzard, with slim wings and fairly long, deeply forked tail. Wing span 145—155 cm. When gliding, further distinguished by white patch behind black tip of underside of wings, which are characteristically angled. Adults speckled rusty brown, with light head, juveniles darker, with browner head. Legs yellow, eyes orange yellow. Downy plumage pale reddish, with white head. Voice: sharp 'heea' and trilling mating calls 'heea-hi-hi-hi-hi-heea'. Rare summer bird in wooded plains and hills. Breeding season: IV—VI. Nest: near top of tree, made of broken branches and brushwood, shallow cup lined with scraps of vegetation, paper and rags. Eggs: 3, chalky white with grey and brown spots, incubated about 4 weeks by ♀. Care of nestlings about 50 days. Diet: small vertebrates, a lot of carrion. Partial migrant. Migration: IX—X and III—IV.

2 **BLACK KITE** *(Milvus migrans)* Size of buzzard, distinguished from Red Kite by only slightly forked tail and dark brown plumage. When tail is spread, fork can hardly be seen, but length of tail and slim wings are good distinguishing characteristics (cf. Buzzard). Adults have dark brown back, grey head and rust brown underside, juveniles darker head and speckled back. Legs yellow, eyes light brown. Downy plumage smoky brown, with yellowish brown underside. Voice: melodious trills 'queeū-kiki-kiki-kik'. Somewhat rare summer bird in woods in wide river valleys and lakeside woods. Breeding season: IV—V. Nest: high up in tree, made of brushwood stuck together with clay, decayed wood, etc. Eggs: 2—3, brown-speckled, incubated 32 days by both parents. Care of nestlings about 6 weeks. Diet: small vertebrates, carrion, dead fish. Migrant. Migration: VIII—IX and III—IV.

3 **MARSH HARRIER** *(Circus aeruginosus)* Large harrier. Wing span 110—140 cm. Distinguishable from buzzard of same size by slimmer body, longer tail and V-position of wings when gliding. ♂ has bright blue-green wing patches and silvery grey tail; ♀ dark brown, with sharply defined light head and nape. Juvenile plumage like ♀, but has light-edged dorsal feathers and often dark head. Legs slender and yellow, eyes yellowish brown. Downy plumage greyish yellow. Voice: utters high 'kweeōō' and whistling 'kwih' mating calls. Nests in large reed-beds, lives near water, roosts in sedge. Breeding season: IV—VI. Nest: hidden in dense rushes, on bent sedge blades over water, made of rushes, sedge and brushwood. Eggs: 4—5, rounded, white, incubated 32—33 days by ♀. Diet: small vertebrates and large insects. Partial migrant. Migration: VIII—X and III—IV.

1

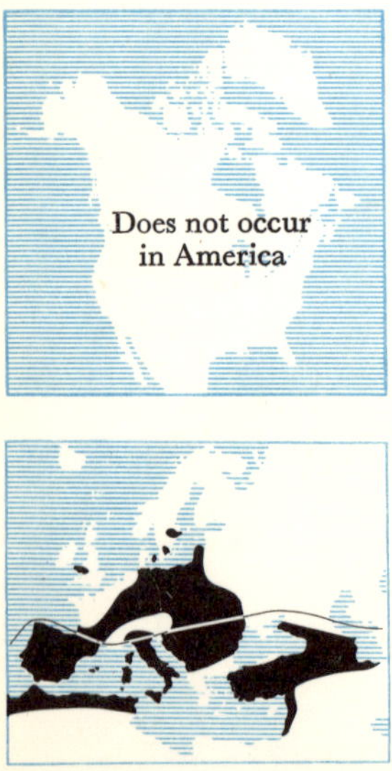

2

3

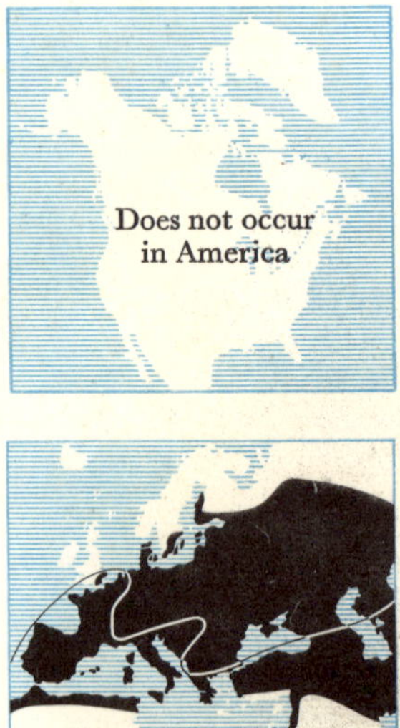

1 ad
1 juv
2 juv
1
2 ad
2
3 ♀
3 ♂

1 **HEN HARRIER - MARSH HAWK** *(Circus cyaneus)* Smaller and slimmer than Marsh Harrier, with narrower wings and white rump. ♂ light grey, with black-tipped wings, ♀ and juvenile dark brown, with brown stripes down light underside. Resembles Montagu's and Pallid Harrier. Legs and eyes yellow. Downy plumage white, with yellow-grey back and dark-ringed eyes. Voice: 'ke-ke-ke-ke-ke' warning calls from nest. Rare summer bird on heaths and moors and in large cultivated areas; not uncommon on fallow land and in woods. Regular winter bird in fields and meadows. Breeding season: V—VI. Nest: on ground, made of stalks, brushwood and heather. Eggs: 4—5, round, incubated 29—30 days by ♀. Care of young 5—6 weeks. Diet: chiefly fieldmice, also other vertebrates and large insects. Pursues prey close to ground, fluttering and gliding with wings held V-wise. Partial migrant. Migration: IX—X and IV.

2 **MONTAGU'S HARRIER** *(Circus pygargus)* Resembles Hen Harrier, but smaller and slimmer, wings more pointed. Flits skilfully after prey. ♂ grey, wings black-tipped, with black band on posterior margin, underside brown-streaked. ♀ usually distinguishable from Hen Harrier by less white rump, otherwise, like juvenile, very similar. Legs and eyes yellow. Downy plumage like Hen Harrier's. Voice: shrill 'kekekeke' calls from nest. Mostly rare nester in bogs and swamps, flood areas and large, wet meadows. Breeding season: V—VI. Nest: on ground, like Hen Harrier's. Eggs: 4—5, round, white, incubated 28—29 days by ♀. Care of young about 5 weeks. Diet and hunting habits like Hen Harrier's. Migrant. Migration: VIII—IX and IV—V.

3 **PALLID HARRIER** *(Circus macrourus)* Slightly slimmer than Hen Harrier, otherwise very similar (both sexes). ♂ seagull grey with very light head and white underside. ♀ like Hen Harrier, but with off-white rump. Juvenile has plain rust brown underside. Cackling voice, ♀ calls 'preepripripri'. Inhabits treeless country, breeds in eastern Europe and Asia, appears in northern, central and western Europe as irruptive migrant, occasionally to nest. Frequents meadows and fields. Breeding, hunting and eating habits like Hen Harrier's. Migrant. Migration: VIII—X and IV—V.

1

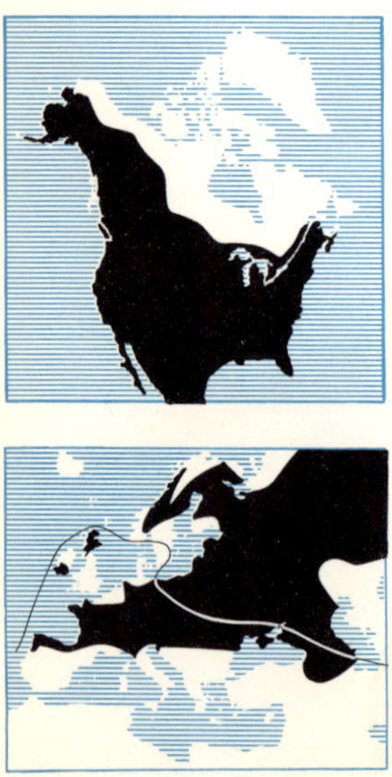

2

3

1 ♂
1 juv
1
2 ♂
2 juv
2
3 ♂
3 juv
3

Order: Birds of Prey - Falconiformes **Family: Ospreys -** Pandionidae

1 **OSPREY** *(Pandion haliaëtus)* Larger than buzzard. Wing span 150—170 cm. Distinguished by pure white underside, white head and broad eye stripe. Long wings, held bent, have black spot on carpal joint. ♂ = ♀. Juvenile has light-edged dorsal feathers. Downy plumage whitish, with dark eye stripe. Voice: 'tchip-tchip-tchi-cheek' mating calls. Seen near inland waters during migration. Nests beside large lakes and ponds. Breeding season: IV—VI. Nest: large, made of sticks and brushwood on tall tree. Fish remains usually found under nest. Eggs: 2—4, yellowish white, grey- and brown-speckled, incubated 35 days by ♂ and ♀. Diet: only fish, usually detected by hovering, caught in vertical dive. Migrant. Migration: VIII—X and IV—V.

Order: Birds of Prey - Falconiformes **Family: Falcons -** Falconidae

2 **SAKER FALCON** *(Falco cherrug)* Larger than Peregrine Falcon, with relatively longer tail and wider wings, otherwise very similar in flight. Distinguished by whitish crown and nape with fine, dark brown streaks, indistinct, thin moustachial stripe and rust brown back. Underside (adults and juvenile) whitish, with rows of vivid brown spots. Legs bluish grey (juvenile) or yellow (adults), eyes dark brown. Downy plumage white. Voice: 'kjikjikji' or drawled 'kyack-kyack'. Inhabits riverside woods and open plains, often nests also on rocks. Breeding season: IV—V. Nest: high up on deciduous tree, often on bare rock. Eggs: 4—5, with red-brown spots, incubated 28—30 days by ♀. Care of young 6—7 weeks. Diet: mainly small mammals, chiefly ground squirrels, also birds up to pigeon and partridge size. Partial migrant, in western Europe accidental bird. Migration: X—XI and III.

3 **PEREGRINE FALCON** *(Falco peregrinus)* About size of crow. Wing span 85—115 cm. ♀ considerably larger than ♂. Adults have dark grey back, densely cross-striped belly and wide, black moustachial stripe. Juvenile has light-edged dorsal feathers and stripes running down belly. Downy plumage white. Distinguished in flight by pointed, bent wings and wedge-shaped, tapering tail. Voice: long 'kek-kek-kek-kek' or 'kjikjikji' warning calls from nest. Rare summer bird in open, rocky country in plains and mountains. In winter seen also in agricultural areas with few trees. Breeding season: III—IV. Lively nuptial flights near nest. Nest: usually in hole or niche in steep rock, without bedding. Eggs: 3—4, thickly brown-speckled, incubated 28—29 days. Care of young about 40 days Diet: birds caught on wing. Partial migrant. Migration: X—XII and II—III.

1

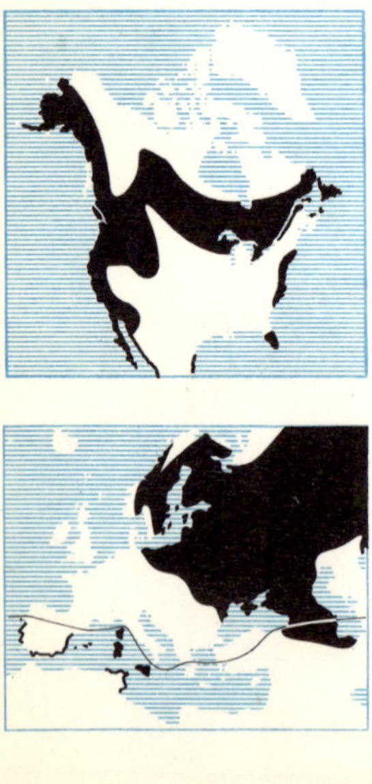

2

3

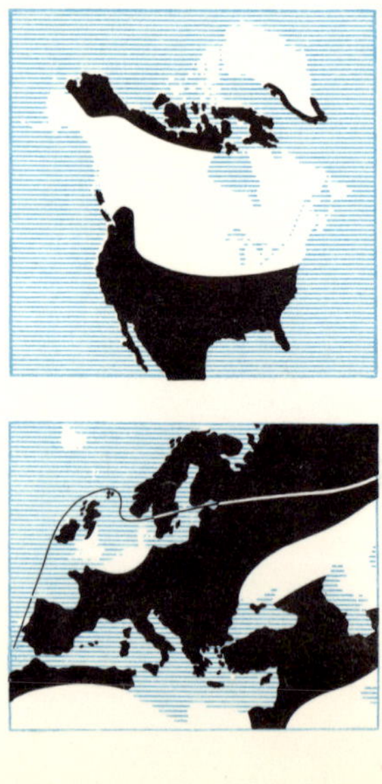

1
1
2 ad
2 juv
3 juv
3 ♂
3 ♀
3
2

1 **HOBBY** *(Falco subbuteo)* Size of pigeon, resembles Peregrine Falcon in flight. Back almost black, underside vividly striped, with russet tibial feathers and under tail-coverts and bold moustachial stripe. Legs and orbital ring yellow, eyes blackish brown. ♂ = ♀. Juvenile has brown-edged dorsal feathers. Downy plumage white. Does not hover, catches prey on wing. Voice: serial 'kik-kik-kik' calls (like Kestrel). Widespread, but not common, nester in copse type woods, hunts in open country, often over ponds. Breeding season: V—VI. Nest: high up on tree; nests of other birds of prey or crows often used. Eggs: 3, with dense red- or yellow-brown speckling, incubated 28 days by ♂ and ♀. Care of young 4 weeks. Diet: small birds, especially swallows and larks, and insects (e.g. dragonflies). Migrant, found in nesting area only from V to IX.

2 **MERLIN - PIGEON HAWK** *(Falco columbarius)* Smallest falcon, often looks like swallow in flight. Wing span 60—70 cm. ♂ has blue-grey back, slate grey, black-edged tail and light brown, streaked underside. ♀ and juvenile have dark-brown back and banded tail. Legs yellow, eyes dark brown. Downy plumage white. Voice: 'kik-kik-kik', like Peregrine Falcon. Nests in bare tundra and high moors in Scandinavia, in winter found in open agricultural country. Breeding season: V—VI. Nest: usually on ground. Eggs: 4—5, with thick brown speckling, incubated 26—30 days by ♂ and ♀. Care of young 25—27 days. Diet: mostly small birds up to size of thrush, occasionally insects and mice. Usually catches flying prey close to ground, sometimes hovers. Partial migrant; regular, though not common winter bird in western and central Europe. Migration: IX—XI and III—IV.

3 **KESTREL** *(Falco tinnunculus)* Small, long-tailed falcon. Wing span 70—80 cm. Most easily identified by lengthy hovering 10—20 m above ground. ♂ has light grey head and tail and red-brown back with dark 'droplet' spots, ♀ and juvenile rust brown back with dark bands. Distinguished in flight by narrow, pointed wings and long, rounded tail. Downy plumage white. Voice: shrill 'kik-kik-kik' and vibrating 'wrreee' calls. Frequent nester in open country with groups of trees, in copses, on rocks, in quarries, in old ruins and on church towers. Breeding season: IV—V. Nest: high up on tree (crow or pigeon nests mostly used) or on bare rock or masonry. Eggs: 5—6, brown-speckled, incubated 28—30 days. Care of young 27—33 days. Diet: chiefly fieldmice. Partial migrant. Migration: IV and IX—X; many birds overwinter.

1

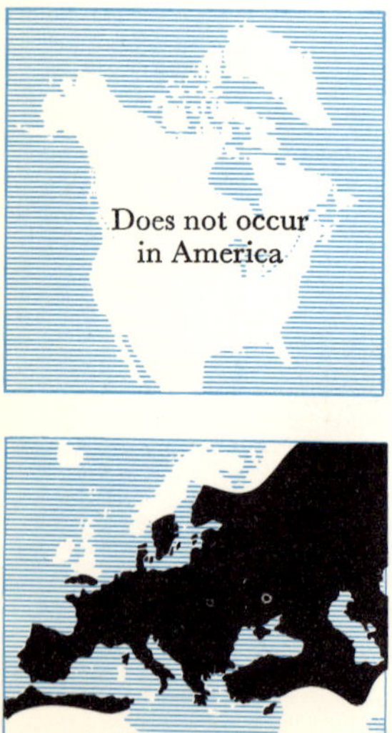

2

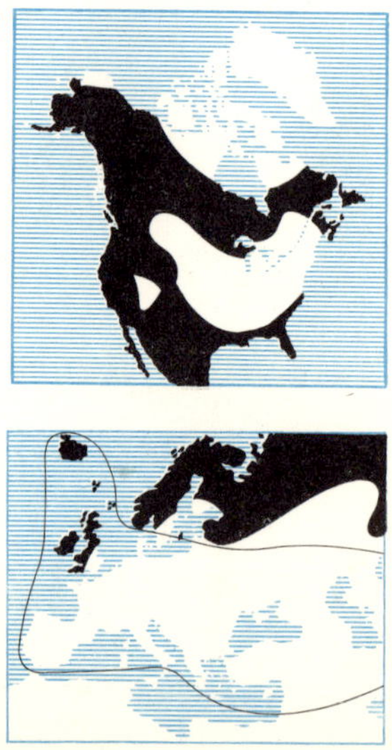

3

1 ad
1 juv
1
2 ♂
2 ♀
2 ♂
3 ♀
3 juv
3 ♂
3 ♂

Order: Birds of Prey - Falconiformes **Family: Falcons** - Falconidae

1 **RED-FOOTED FALCON** *(Falco vespertinus)* Size of Kestrel. In flight, underside of wings slate grey to black (♂), or thickly banded in black and white (♀ and juv.). ♂ has coral-red legs, cere and orbital ring, ♀ grey, black-banded back and tail and rusty yellow crown and nape, juvenile brownish back, striped underside and yellow legs. Downy plumage white. Voice: 'ki ki ki ki'. East European species, nests in flat, open country with groups of trees. Seen irregularly in central and western Europe when migrating, in some years as irruptive bird. Breeding season: V—VI. Nest: on tree, usually with others. Eggs: 4—5, brown-speckled, incubated 22—23 days by ♂ and ♀. Care of young 26—28 days. Diet: almost entirely insects (beetles, grasshoppers, dragonflies). Migrant. Migration: VIII—IX and IV—V.

Order: Game Birds - Galliformes **Family: Grouse** - Tetraonidae

2 **WILLOW GROUSE - WILLOW PTARMIGAN** *(Lagopus lagopus)* Larger than Partridge. Legs and feet feathered to tips, especially noticeable in winter plumage. In summer dark reddish brown, with white wings, in winter pure white with black marks on wings. Subspecies inhabiting British Isles, Red Grouse (2 a), stays brown in winter. Closely resembles Ptarmigan. In transitional plumage variegated. ♂ = ♀ = juv. Flies fast, with loudly whirring wings. Downy plumage yellowish brown, with distinct black spots on head and back. Voice: 'err-rek, ok-ok-ok'. Inhabits moors and heaths. Breeding season: V—VI. Nest: under bush or in heather. Eggs: 8—12, ochre yellow, incubated 21—24 days by ♀. Chicks remain with adults in large coveys. Diet: berries, willow and birch buds, etc. Resident bird.

3 **PTARMIGAN - ROCK PTARMIGAN** *(Lagopus mutus)* Larger than Partridge. In summer, ♂ has black-spotted, grey-brown to grey back and breast, ♀ yellowish brown, marbled back. Juvenile like ♀, but with grey wings and belly. White wings in all plumages. Distinguished from Willow Grouse by thinner beak and black eye stripe. Flies low, with purring sound. Voice: harsh croaking and squawking calls, 'ārrr', etc. In Alps and Pyrenees nests on stony slopes above forest belt, in north in rocky and mountain tundra. Breeding season: V—VII. Nest: in rubble or bush. Eggs: 6—10, thinly speckled with brown. Care of young as Willow Grouse. Diet: various Alpine and tundra plants, some insects (especially liked by young). Resident bird.

1

2

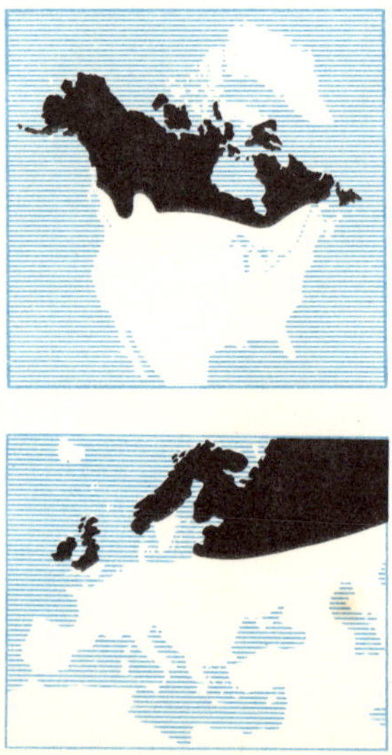

3

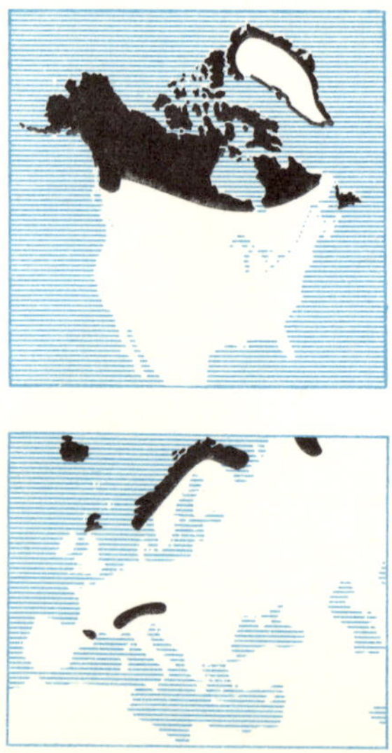

1 juv
1 ♂
1 ♀
2 ♂ W
2a ♂ S
2 ♂ S
2 a
3 ♂ W
3 ♂ S
3 ♀ S
3 ♂ S

Order: Game Birds - Galliformes **Family: Grouse -** Tetraonidae

1 **BLACK GROUSE** *(Lyrurus tetrix)* Size of domestic hen. ♂ in nuptial plumage unmistakable, with gleaming black-blue plumage, lyre-shaped tail and brilliant white under tail-coverts. ♀ rust brown, with dark bands and slightly forked tail. Juvenile = ♀. Downy plumage rusty brown, faintly spotted. Voice: courting ♂ utters hissing 'whushee' and rookooing 'krōō krōō krōō' calls; ♀ cackles. Locally widespread nester on heaths and moors and in stunted timber belt in mountains. Mostly gregarious, especially at mating time, on repeatedly revisited courting ('lekking') ground. Breeding season: V—VI. Nest: on ground, among heather or bilberry plants. Eggs: 6—10, ochre yellow, with small, red- or black-brown blotches and dots, incubated 25—27 days by ♀. Young fledge at an early age, follow parents about 3 weeks. Diet: mainly vegetarian (buds, shoots, berries), sometimes insects. Resident bird.

2 **CAPERCAILLIE** *(Tetrao urogallus)* Forest bird almost as large as turkey, with thick, hooked beak. ♂ black and dark brown, with white spots on anterior border of wings. ♀ about $\frac{1}{3}$ smaller, rusty brown. Distinguished from Black Grouse by size and rounded tail. Juvenile resembles ♀. Downy plumage yellower on head than in Black Grouse, otherwise similar. Voice: courting ♂ utters wooden tapping and grinding sounds, ♀ cackles. Clatters up into air if flushed. Rare nester in large mountain forests, especially pine-woods. Breeding season: IV—VI. Nest: on ground, hidden in heather or bushes. Eggs: 6—10, size of hen's egg, ochre yellow with dark brown spots, incubated 26—28 days by ♀. Young remain with ♀ until winter. Diet: mainly vegetarian (needles, shoots, buds). Resident bird.

3 **HAZEL HEN** *(Tetrastes bonasia)* Size of Partridge, with rusty brown and grey, black and white-speckled back and rounded grey tail edged with broad band. ♂ has short crest and black throat patch framed with white; ♀ has white throat. Juvenile resembles ♀. Downy plumage plain rusty brown, with black eye stripe. Whirring flight. Voice: unusually high and shrill, courting ♂ calls 'trissi-tseri-tsi, tsi-tsui'. Inhabits lonely woods with abundant undergrowth, especially in mountains. Breeding season: IV—V. Nest: thinly lined depression in ground, hidden by vegetation. Eggs: 8—10, reddish yellow, with a few brown spots, incubated 21—25 days by ♀. Young follow ♀ (often ♂ and ♀) up to 3 months. Diet: mainly deciduous tree buds and shoots, also berries, insects and snails. Resident bird; in north partial migrant.

1

2

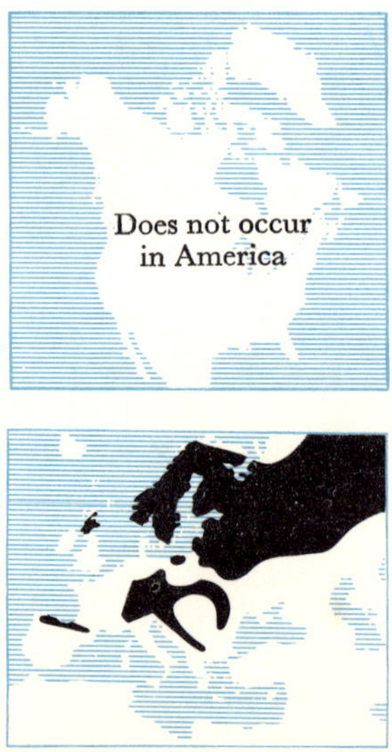

3

1 ♂
1 ♀
1 ♂ display
2 ♂
2 ♂
2 ♀
2 ♂ display
3 ♀
3 ♂
3 ♂

Order: Game Birds - Galliformes **Family: Pheasants, Partridges and Quails** - Phasianidae

1 **ROCK PARTRIDGE - CHUKAR** *(Alectoris graeca)* Resembles Partridge in form, size and flight. Adult has white, black-framed throat, grey back and black- and white-banded sides. Beak and legs red (unlike Partridge, but like Red-legged Partridge). Juvenile has 'droplet' spots on back and breast. Downy plumage red-brown on back, with black and white markings. Voice: mating call of ♂ loud 'kakelik'. If flushed, clatters up into air, uttering shrill 'rittchi rittchi'. Inhabits warm, bare slopes with rock and boulders, mostly above tree belt in mountains. Breeding season: V—VII. Nest: small depression in ground. Eggs: 9—14, with red-brown spots, incubated 24—26 days by ♀. Young follow parents until autumn. Diet: buds, leaves, berries, seeds, some insects. Resident bird.

2 **RED-LEGGED PARTRIDGE** *(Alectoris rufa)* Closely resembles Rock Partridge in form and behaviour, but has purple brown back and black frame of throat patch fans out into short stripes. Juvenile's throat patch is unframed. Downy plumage spotted on back, as in Partridge. With reddish brown back and bright red-brown tail, resembles Partridge in flight, but distinguished by colouring of throat and underside. Voice: mating call of ♂ loud 'tschreck, chuker', when flushed 'kuk-kuk'. Nests on dry, sandy or stony ground on low mountain slopes and in vineyards and cornfields. Breeding season: V. Nest, eggs and care of young like Rock Partridge, eggs incubated 23—24 days. Diet: mainly plant parts and seeds, also insects, molluscs and worms. Resident bird.

3 **PARTRIDGE - GRAY PARTRIDGE** *(Perdix perdix)* Stocky, short-tailed bird with rounded wings. ♂ has brown horseshoe 'breastplate'. Downy plumage reddish brown with black spots, underside light yellow. Alternates audible whirring flight with gliding; looks yellow from below in flight. Voice: loud 'kirr-ic', when flushed 'krikrikri'. Inhabits cultivated country at low and moderate altitudes, with corn, clover and beet fields. Breeding season: V—VI. Nest: shallow depression lined with a few grass blades, well hidden under bush and dense plants. Eggs: 10—20, plain greenish grey, incubated 24—25 days by ♀. Young led by ♂ and ♀, spend winter as covey. Diet: seeds, green parts of plants, in summer mainly insects and worms. Resident bird.

1

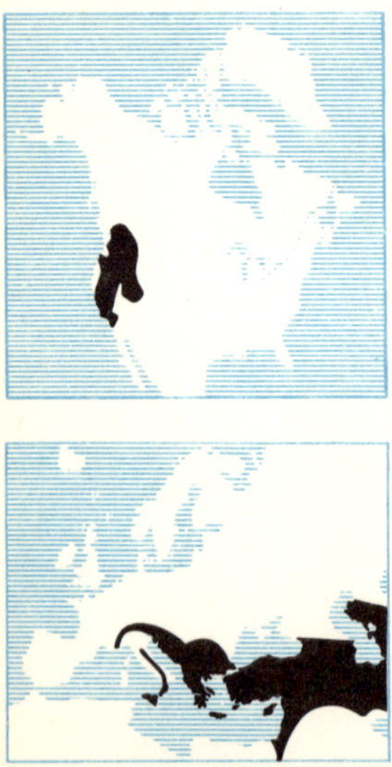

2

3

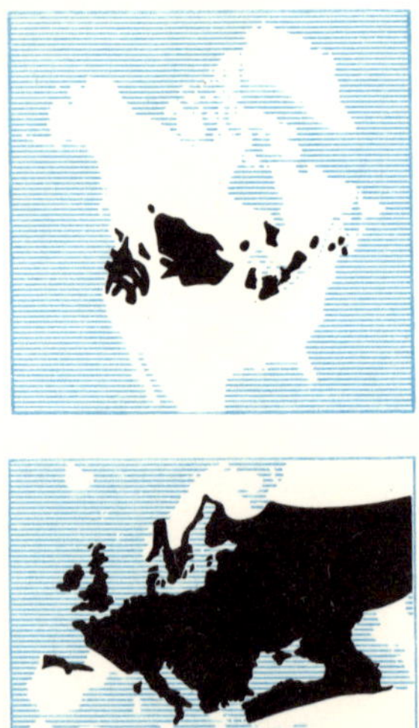

1
1
2
2
3 ♂
3 ♀
3 ♂

Order: Game Birds - Galliformes **Family: Pheasants, Partridges and Quails** - Phasianidae

1 **QUAIL** *(Coturnix coturnix)* Size of Blackbird, with short tail and small head. Earthen brown, with dark and white streaks. ♂ has black markings on throat (not ♀ or juvenile). Downy plumage rust brown, with 2 sharp stripes down body. Unmistakable in flight because of small size, but hard to flush. Voice: carrying, rhythmical 'quic-quic-ic', also heard during flight and at night. Secretive, inhabits cultivated fields and meadows. Breeding season: VI—VII. Nest: in small depression in ground. Eggs: 7—14, brownish yellow, with dense dark brown speckles and spots, incubated about 17 days by ♀. Young also led by ♀, not more than 1 month. Diet: mainly different weed seeds and parts of leaves, in summer insects. Migrant. Migration: IV—V and IX—X.

2 **PHEASANT - RING-NECKED PHEASANT** *(Phasianus colchicus)* Size of domestic fowl, with long, tapering tail. Plumage of ♂ has metallic lustre, head dark green with bright red skin flaps (wattles) and usually white neck-band. ♀ completely earthen brown, with dark-spotted back and shorter tail. Juvenile like ♀. Downy plumage russet, with black spots on back. Flies up noisily when flushed. Voice: mating call of ♂ loud 'korrk-kok', followed by flapping of wings; birds retiring to tree to sleep call 'kut kut kuttuc'. Common nester in cultivated country with trees and bushes. Often gregarious. Breeding season: V—VI. Nest: sparsely lined depression in ground. Eggs: 10—12, plain greyish green, incubated 24—25 days by ♀. Young led 1 month by ♀. Diet: in summer mainly insects, worms and molluscs, otherwise largely seeds and greenstuff. Resident bird.

Order: Rails and Allies - Gruiformes **Family: Cranes** - Gruidae

3 **CRANE** *(Grus grus)* Larger than stork, distinguished from herons by dropping tail feathers. ♂ = ♀. Juvenile plumage browner, with no colourful head markings and no plumes on back and tail. Flies with neck and legs extended, if in numbers in echelon formation. Voice: resonant 'kroo, krerr'. Rare and retiring summer bird in large swamps, seen in flocks in fields and meadows near water during migration. Breeding season: IV—V. Nest: large heap of sedge in shallow water. Eggs: 2, large, greenish grey with a few spots, incubated 29—30 days by ♂ and ♀. Young fledged in about 10 weeks. Diet: seeds, berries, maize, plus insects and small vertebrates. Migrant. Migration: III—V and IX—X.

1

2
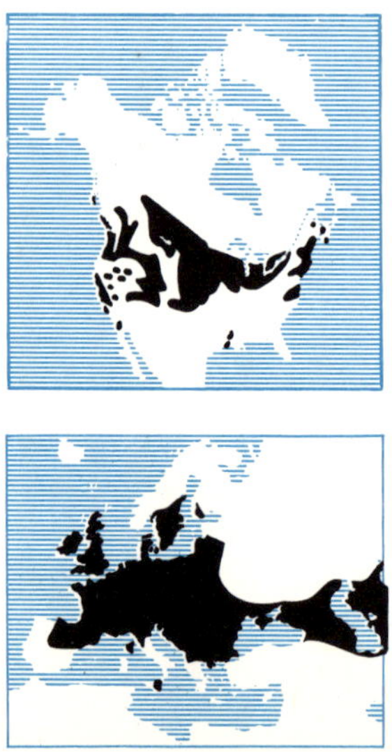

3

1 ♂
1
♀
2 ♀
2 ♂ *torquatus*
2 ♂ *colchicus*
3
3 juv
3 ad
3 D

Order: Rails and Allies - Gruiformes **Family: Bustards** - Otididae

1 **GREAT BUSTARD** *(Otis tarda)* Like turkey in size and form, with copper brown back, white underside, thick, light grey neck and strong legs. ♂ has whiskers at base of beak, ♀ much smaller, with no whiskers. Juvenile duller, with rusty crown. Downy plumage cream, with black spots and stripes. Voice: snorts and hisses, seldom audible, accompany bizarre courting display of ♂. Rare inhabitant of thinly populated, treeless, dry cultivated steppes; mostly gregarious, but very timid. Breeding season: V—VI. Nest: on ground, unlined. Eggs: 2, grey-green with a few brown spots, incubated about 25 days by ♀. Nidifugous young led 6—7 weeks, only by ♀. Diet: mainly leaves, buds, seeds, also insects, lizards, fieldmice. Resident bird or partial migrant.

2 **LITTLE BUSTARD** *(Otis tetrax)* Size of domestic fowl. Sandy, with fine black ripple marks. ♂ in nuptial plumage has black and white collar, ♀ striped and spotted breast and sides. Juv = ♀. Downy plumage somewhat ligher than Great Bustard's. In flight appears almost white, because of large white areas on wings and white underside. Voice: display note of ♂ a loud, throbbing 'ptrrr'; utters whistling sounds when flying. Rare steppe-dweller, found in open plains with large corn- and clover-fields. Breeding season: V—VI. Nest: hollow in ground, lined with dry grass stems. Eggs: 3—4, olive green with a few brown spots, incubated 20—21 days by ♀. Young led about 5 weeks by both parents. Diet: as for Great Bustard. Partial migrant.

Order: Rails and Allies - Gruiformes **Family: Crakes, Rails and Coots** - Rallidae

3 **COOT** *(Fulica atra)* Black aquatic bird almost size of duck, with white beak and frontal shield. Eyes red, clumsy-looking feet have lobed webbing on long toes. ♂ = ♀. Juvenile has greyish white throat and belly. Downy plumage black, with orange red markings on head and neck. Escapes from danger flapping wings and beating water loudly with feet. Dives laboriously, for short periods only. Common summer birds on stagnant inland waters, in winter found in large flocks on slow-flowing ice-free rivers and lakes. Breeding season: IV—V, 1—2 broods a year. Nest: pile of broken rushes in waterside reeds. Eggs: 6—9, yellowish white, densely speckled with black, incubated 21—22 days. Diet: subaquatic plants, shoots of waterside plants, insects, molluscs. Partial migrant. Migration: III and X—XI.

1

2
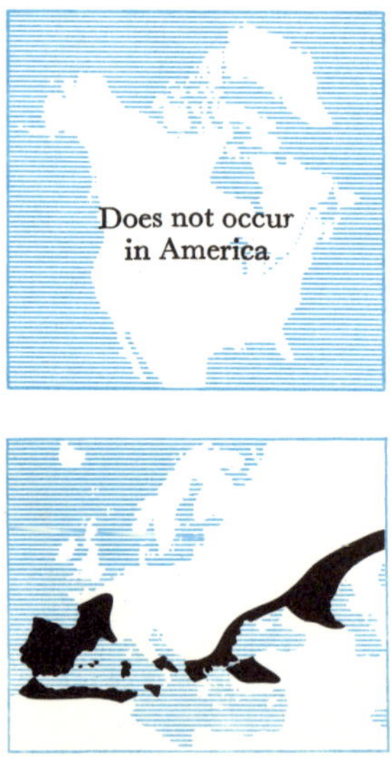

3

display
1 ♂
1 ♂
1 ♀
2 ♂
2 ♀
3 juv
3 ad
3 D

Order: Rails and Allies - Gruiformes **Family: Crakes, Rails and Coots** - Rallidae

1 **MOORHEN - COMMON GALLINULE** *(Gallinula chloropus)* Slightly smaller than Partridge. Adults dark, with red frontal shield, red eyes, short tail with gleaming white under tail-coverts and green legs. Juvenile grey-brown. Downy plumage black, with red beak. Sits high in water when swimming, nods head rhythmically and jerks erect tail. Voice: regular explosive 'kittick' and excited 'kik-kik-kik'. Common on stagnant water with dense waterside vegetation and water-weeds, in winter found on ice-free rivers. Breeding season: IV—VII, 2 (sometimes 3) broods a year. Nest: basin made of reed-grass leaves and blades near water. Eggs: 6—10, yellow, with brown to black speckles and spots, incubated 19—22 days by both parents. Diet: various small aquatic animals and green parts of plants. Partial migrant. Migration: III—IV and IX—XI.

2 **WATER RAIL** *(Rallus aquaticus)* Size of Blackbird, with short tail and long, thin beak. Eyes and beak red, legs brown. ♂ = ♀. Juvenile mainly brown. Downy plumage deep black, with pale legs and beak. Fairly common, seldom seen, but often heard. Voice: very loquacious; explosive 'kik-kik' calls and frequent squeaky 'krooihf' followed by low groaning. Nests at edge of thickly overgrown swamps and marshy ponds. Breeding season: IV—V and VIII, probably 2 broods a year. Nest: made of reed, rush or sedge leaves in thick swamp vegetation, usually under cover. Eggs: 6—12, creamy yellow with a few brown spots, incubated 19—21 days by both parents. Young nidifugous, independent at about 8 weeks. Diet: various animals, a few plants and seeds. Partial migrant. Migration: III—V and IX—XI.

3 **CORNCRAKE** *(Crex crex)* Same size as Water Rail, with short beak. Light brown, with russet wings (conspicuous in flight). ♂ = ♀ = juv. Downy plumage black, with blackish brown beak. Voice: persistently repeated 'ārp-ārp' heard chiefly after dusk and at night. Hides in grass, seldom flies. Inhabits damp mowing meadows, especially near rivers, and clover and lucerne fields, also found in mountains at low altitudes. Becoming rarer everywhere. Breeding season: VI—VII. Nest: hollow in ground lined with grass leaves and blades. Eggs: 8—12, yellow with grey- to red-brown spots, incubated about 19 days by ♀. Young led about 5 weeks, only by ♀. Diet: insects, spiders, snails, slugs, various seeds. Migrant. Migration: V and IX—X.

1

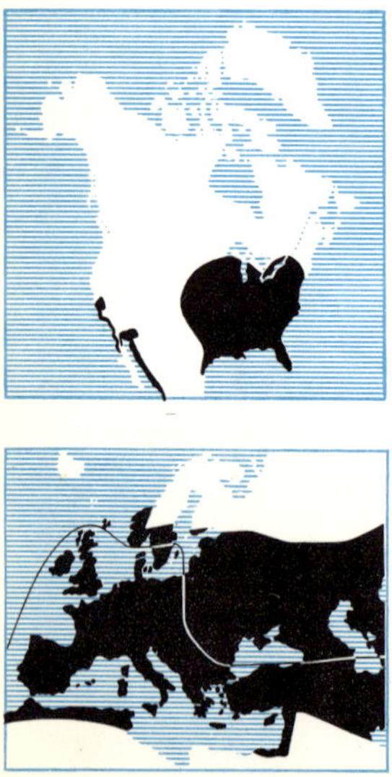

2

3

1
1 D
1 juv
2 ad
2
2 D
2 juv
3 ad
3
3 juv

Order: Rails and Allies - Gruiformes **Family: Crakes, Rails and Coots** - Rallidae

1 **SPOTTED CRAKE** *(Porzana porzana)* About size of Water Rail, with short beak and white-speckled back and breast. Juvenile brown, with whitish underside. Downy plumage deep black. Seldom flies, lives secretively in thick reeds and sedge beside marshy lakes and ponds, in wet meadows, near dykes, etc. Voice: whistling 'whitt', heard after dusk and at night, persistently repeated at intervals of about 1½ seconds, often for hours. Breeding season: V—VII, 2 broods a year. Nest: deep cup made of reed-grass and rush blades on bent sedge in thickest marsh vegetation. Care of brood as for Water Rail. Diet: insects, molluscs, worms, seeds. Migrant. Migration: III—V and VIII—X.

2 **LITTLE CRAKE** *(Porzana parva)* About size of Starling, with short beak, red at base, and short, erect tail. ♂ has brown back and blue-grey underside, ♀ is nut brown. Juvenile = ♀, but has white-spotted back. Legs green. Downy plumage black, with pale yellow beak. Very retiring bird. Voice: descending display call of ♂ 'goot-goot-gootgootgoot', heard most frequently, also calls 'kick-körr'. Nests in swamps, bogs and in dense vegetation beside marshy ponds. Breeding season: V—VI, usually 2 broods a year. Nest: deep, made of sedge and reed-grass leaves just over water. Eggs: usually 6—8, greyish yellow with faded red-brown spots, incubated 20—21 days by both parents. Young led 4—5 weeks by both parents. Diet: insects, spiders, molluscs, occasionally seeds. Migrant, some partial migrants. Migration: IV and VIII—IX.

3 **BAILLON'S CRAKE** *(Porzana pusilla)* Slightly larger than Sparrow. Distinguished from Little Crake by white-spotted back, vividly banded sides, brownish-red legs and absence of red mark on beak. (Water Rail and Spotted Crake also have banded sides, but former has long beak and latter browner, more speckled breast.) ♂ = ♀. Juvenile has whitish and brown underside. Downy plumage black, with ivory-white beak. Voice: explosive, rolling 'kit-kit-kuttt'. Rare inhabitant of marshy ground beside marshy ponds, lakes and creeks. Breeding season: V—VII, 1 brood. Nest: deep cup made of sedge leaves, usually in clump of sedge surrounded by water, covered over. Eggs: 6—8, yellowish-brown, with violet and brown spots, incubated 20—21 days by both parents. Care of brood, and diet as for Little Crake. Migrant. Migration: IV and IX—X.

1

2

3

1
1 ad
1 juv
1 D
2
2 ♀
2 ♂
3 juv
3 ad

Order: Waders - Charadriiformes **Family: Oystercatchers** - Haematopodidae

1 **OYSTERCATCHER - AMERICAN OYSTERCATCHER** *(Haematopus ostralegus)* Almost as large as crow. Recognizable by black and white plumage, bright red beak and red legs. Juvenile plumage has light scales on back. Downy plumage yellowish grey, with narrow black streaks on back. In flight, further distinguished by white rump, broad white areas on wings and wide black band at end of tail. Voice: loud, melodious, 'klee-eep', at nesting site incessant, vociferous 'kip-kip-kip'. Widespread inhabitant of coasts and dry coastal grasslands, occasionally seen on inland waters during migration. Breeding season: V—VI. Nest: shallow depression in sand or pebbles, lined with broken shells and small stones. Eggs: 3—4, sandy, with grey and black spots, incubated about 27 days by both parents. Diet: sea worms, mussels, crabs. Partial migrant. Migration: III—IV and VIII—IX

Order: Waders - Charadriiformes **Family: Plovers and Lapwings** - Charadriidae

2 **LAPWING** *(Vanellus vanellus)* Size of pigeon, easily recognized by opalescent sheen on black-green dorsal plumage and erectile, pointed crest. ♀ has white-spotted breast band. Juvenile has pale scaly dorsal plumage. Downy plumage brownish grey. Flies rather slowly and jerkily, identifiable by rounded wings, white, black-ended tail and cinnamon under tail-coverts. Voice: loud 'pee-wit', display call 'peerrweet-weet-weet'. Common nester in wet meadows and cultivated fields. Breeding season: III—V, 1 brood a year. Nest: sparsely lined with straw and grass leaves. Eggs: 4, top-shaped, olive green with thick, brown-black spots, incubated 24—25 days by both parents. Diet: insects, spiders, earthworms, small snails. Partial migrant. Migration: II—III and VIII—X.

3 **TURNSTONE - RUDDY TURNSTONE** *(Arenaria interpres)* About size of Blackbird, with short legs and tail, red-brown back, white underside and broad black breast band. ♂ = ♀. Winter plumage grey. Juvenile has light-edged feathers. With broad white wing bands, white rump with black horseshoe spot and pied head, looks variegated in flight. Voice: high-pitched 'kitititit' and sharp 'tche-wik tche-wi-i-i-i-ck'. Fairly common littoral bird, especially on small, rocky coastal islands, occasionally seen inland when migrating. Breeding season: V—VI. Nest: shallow hollow in gravel, sparsely lined with dry vegetation. Eggs: 4, grey-green, spotted, incubated 23—27 days. Diet: crustaceans, insects, molluscs, preferably gathered from washed-up seaweed. Migrant. Migration: IV—VI and VIII—X.

1
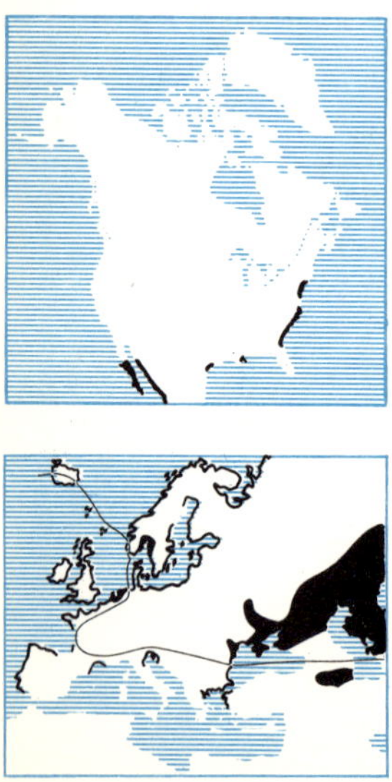

2

3
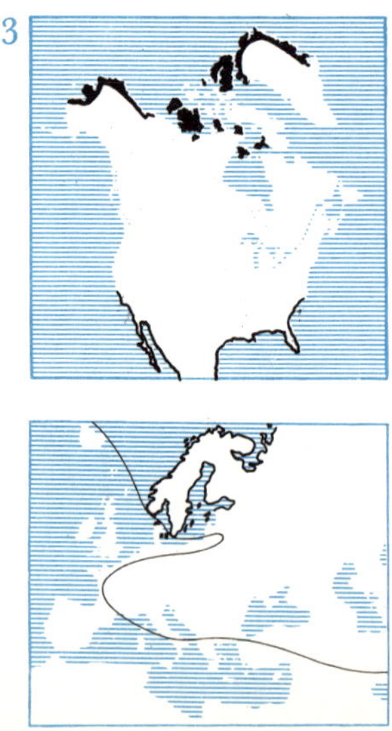

1 S
1
1 D
1 imm
2 D
2
2 ad
2 juv
3 S
3 W

1 **RINGED PLOVER** *(Charadrius hiaticula)* About size of Song Thrush. Distinguished from Little Ringed Plover by size, head markings (black frontal band without white upper border), wider breast band, yellow base of beak, orange yellow legs and, in flight, white band on wings. ♂ = ♀. Juvenile has black beak, narrow breast band and scaly dorsal plumage. Downy plumage grey-brown, with dark crown stripe (absent in Little Ringed Plover). Voice: melodious 'tooi tooi'. Common summer bird on sandy shores, during migration seen inland on drained ponds. Breeding season: V—VII, 2 broods a year. Nest: shallow hollow in sand, usually lined with pebbles and small shells. Eggs: 4, top-shaped, sandy, black-speckled, incubated 23—26 days. Partial migrant. Migration: III—IV and VIII—IX.

2 **LITTLE RINGED PLOVER** *(Charadrius dubius)* Resembles Ringed Plover, but somewhat smaller, with thinner, plain-coloured beak, narrower breast band and white line above black frontal band. ♂ = ♀. Juvenile similar to Ringed Plover, but has no wing bands. Like other plovers, walks with rolling gait, then suddenly stops dead. Voice: high, whistling 'teeū', when flying hoarse 'chreea chreea chreea'. Widespread summer bird on large, gravelly river banks nad sandy lakes and ponds. Breeding season: V—VII, sometimes 2 broods a year. Nest: shallow hollow in sandy ground, lined with small stones or shells. Eggs: 4, top-shaped, sandy, with a few grey and dark brown spots, incubated 22—26 days by both parents. Young led about 22 days. Diet: insects, spiders, worms. Migrant. Migration: IV—V and VIII—IX.

3 **KENTISH PLOVER - SNOWY PLOVER** *(Charadrius alexandrinus)* About size of lark. Black spot on either side of breast (no breast band), black legs and beak. ♂ has reddish brown crown, ♀ grey-brown markings instead of black. Juvenile like Little Ringed Plover, but black legs. Downy plumage like Ringed Plover's. Narrow white wing band visible during flight. Voice: 'wit-wit-wit' or 'kittup' and repeated 'pee-i'. Widespread littoral bird, seldom seen inland during migration. Breeding season: V—VI. Nest: shallow depression in sand. Eggs: 3, sandy, with irregularly distributed black spots and streaks, incubated 24—27 days by both parents. If disturbed, parents draw off intruder by pretending to have broken wing. Diet: insects, worms, molluscs, crustaceans. Migrant and partial migrant. Migration: III—IV and VIII—IX.

1

2

3

1
1 juv
2
2
2 juv
2 D
3
3
3 juv

1 **GREY PLOVER - BLACK-BELLIED PLOVER** *(Pluvialis squatarola)* Slightly smaller than Lapwing. Winter plumage light, with finely speckled grey back and pure white underside. In nuptial plumage, underside deep black (♂) or brownish black (♀), belly and rump white. Beak and legs black. Distinguished from Golden Plover by grey back and, in flight, black axillary spot and white wing bands. Voice: rhythmical call, usually of 3 syllables, 'tlee-oo-ee'. Nests in arctic tundra, winters on European coasts. During migration seen in small numbers inland, on stretches of mud and sand by large ponds and lakes. Breeding season: VI—VII, 1 brood a year. Nest: on sandy ground or turf. Eggs: 4, pear-shaped, greyish yellow with black-brown spots, incubated about 25 days. Diet: crabs, molluscs, worms, insects. Migrant. Migration: IV—VI and VIII—XI.

2 **GOLDEN PLOVER** *(Pluvialis apricaria)* Resembles Grey Plover, but rather smaller, with thinner beak and gold-speckled back. Lacks axillary spots and wing bands and has dark rump. In nuptial plumage has black belly. ♂ = ♀. In juvenile and winter plumage has white underside. Downy plumage yellow brown on back, with black spots and lores, legs bluish green. Voice: fluted 'tlūi'. Nests in wet moorlands, seen on drained ponds and gleaned fields when migrating. Breeding season: IV—V, 1 brood a year. ♂ displays and calls in flight. Nest: shallow depression in moss, thinly lined with grass. Eggs: 4, pear-shaped, greyish yellow, thickly marked with brownish black spots, incubated 27 days by both parents. Young led about 4 weeks. Diet: insects, snails and slugs, worms. Migrant and partial migrant. Migration: III—IV and VIII—IX.

3 **DOTTEREL** *(Eudromias morinellus)* Size of large thrush, mostly dark. In nuptial plumage has long, broad white eye stripe and white breast band. White throat distinctive in flight. ♂ = ♀. Winter and juvenile plumage has brown back with dark 'droplet' spots and streaks, underside white. Beak black, legs yellowish brown. Downy plumage ochre brown with vivid white spots, legs bluish. Voice: trilling 'tit-ri-titi-ri' and repeated 'quee-quee-skir'. Nests in treeless tundra with stony mountain ridges and in similar country in the Alps. Breeding season: VI—VII. Nest: deep hollow in ground, thinly lined with moss and lichen. Eggs: 3, olive brown, thickly marked with blackish brown spots. Incubated 22—25 days and young led about 1 month, both by ♀. Diet: mostly insects and spiders, less often snails and worms. Migrant. Migration: IV—V and IX—X.

1

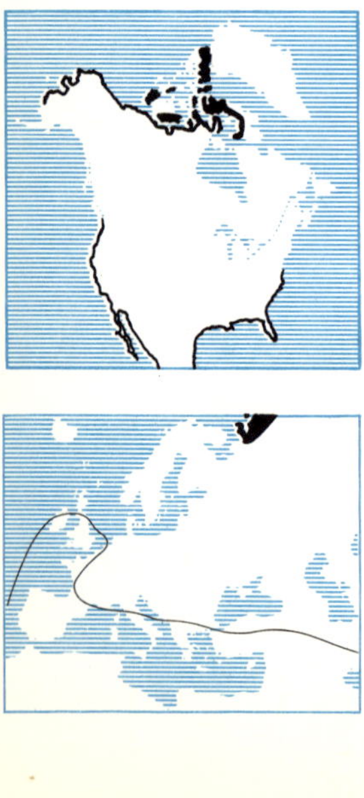

2

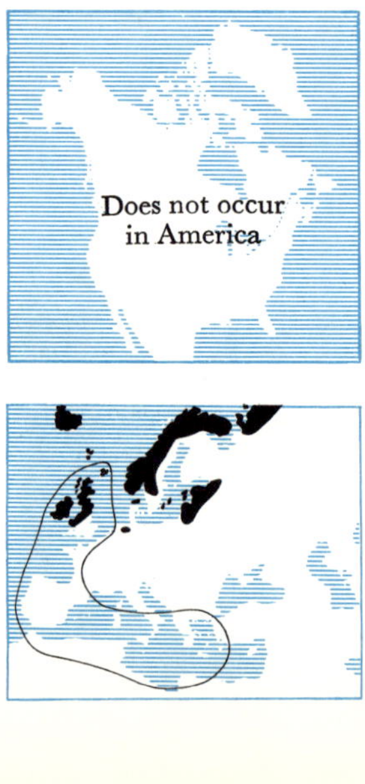

3

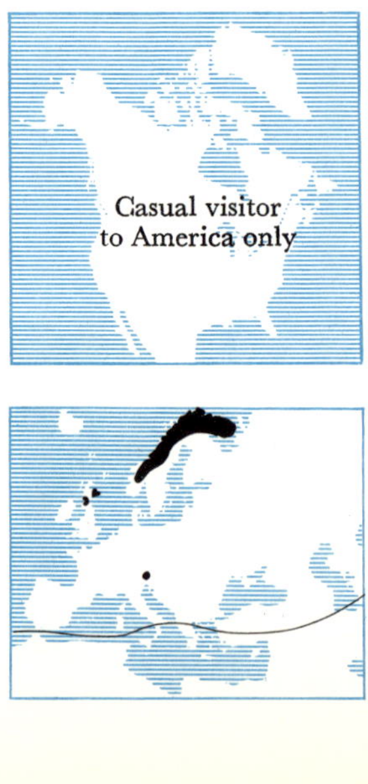

1 W
1 S
2 S
2 S
2 W
3 S
3 S
3 W

Order: Waders - Charadriiformes **Family: Snipes and Sandpipers** - Scolopacidae

1 **SNIPE - COMMON SNIPE** *(Gallinago gallinago)* About size of thrush, with very long, straight beak and longish legs. Back blackish brown, with 2—4 light stripes. ♂ = ♀ = juv. Downy plumage chestnut, with black spots and vivid white dots. When flushed, describes several zigzags and calls 'catch' (as distinct from Great Snipe and Jack Snipe). Characteristic display of ♂♂ repeated plummeting swoops, accompanied by drumming 'huhuhuhuhuhuhu'. Voice (heard mostly on ground): rhythmically repeated 'chic-ka-chic-ka'. Common nester in wet meadows, bogs and swamps, seen during migration on drained ponds. Breeding season: IV—VII, 1 brood a year. Eggs: 4, pear-shaped, olive green, with large, dark spots, incubated 19—21 days by both parents. Partial migrant. Migration: III—IV and IX—XI.

2 **GREAT SNIPE** *(Gallinago media)* Slightly larger and stockier than Snipe, with thinner, shorter beak. ♂ = ♀ = juv. Downy plumage brown, underside white. Flight slower and more ungainly, further distinguished by white-edged tail and light stripe on wings. Flies in straight line, usually in silence. Voice: faint 'bibbelibibib' during display of ♂♂ on ground. Nests in swamps, moors and wet meadows, seen sporadically on ponds when migrating. Breeding season: V—VII. Courtship on ground (not in air, like other snipes). Nest: on ground, hidden, lightly lined with plant material. Eggs: 4, yellowish grey with large, ash-grey and black spots, incubated 20—24 days. Young led about 1 month by both parents. Diet: worms, insect larvae, seeds. Migrant. Migration: IV and VIII—IX.

3 **JACK SNIPE** *(Lymnocryptes minimus)* About size of lark, with shortish beak and short legs. Back almost black, with metallic lustre and 2 wide, light stripes. ♂ = ♀. Juvenile's back dull. Downy plumage chestnut, like Snipe's. Flushed only at close range, flies slowly and in silence; after short, straight flight usually drops again. Distinguished in flight by black back with light V-stripes, dark crown stripe and pointed brown tail with no white feathers. Voice: rhythmical 'lok-toggi, lok-toggi', like trotting horse, during display flight. Inhabits damp moors and swamps, occasionally seen on muddy pond and river banks when migrating. Breeding season: V—VII. Nest, eggs and diet as for Snipe. Migrant. Migration: IV and IX—X.

1

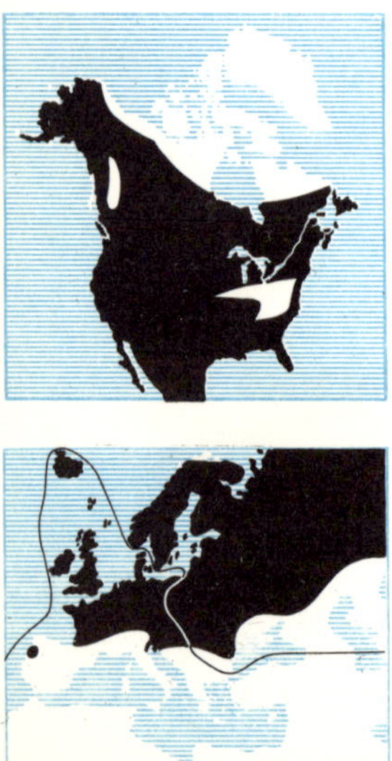

2

3

1
1
1
1 D
2
2
3
3

1 **WOODCOCK** *(Scolopax rusticola)* Almost partridge size, with fairly large head, very long beak, short tail and rather short legs. Brown colouring matches forest floor so closely that bird is seen mostly only when flying. ♂ = ♀ = juv. Downy plumage cinnamon, with large chestnut spots. Display flight (roding) performed low and slowly over woods after dusk, always accompanied by deep, throbbing 'orrrt-orrrt' and high, piercing 'tsiwick' calls. Secretive forest-dweller, nests mainly in mixed woods at low and high altitudes. Breeding season: IV—VII, 1—2 broods a year. Nest: shallow depression in ground lined with dry leaves and moss. Eggs: 4, rounded oval, cream with a few red-brown speckles and spots, incubated 22—24 days by ♀. Diet: chiefly worms and insect larvae. Partial migrant. Migration: III—IV and IX—X.

2 **CURLEW** *(Numenius arquata)* Size of crow, with long legs and long, curved beak. Mostly brown and speckled; rump, belly and underside of wings whitish. Crown unstriped, legs grey. ♂ = ♀. Juvenile more rust-coloured. Downy plumage rusty yellow with large black spots. Voice: carrying, fluty 'cour-li' while flying. Nests in large low-lying meadows and heaths, seen by muddy ponds and river mouths when migrating. Breeding season: IV—V, 1 brood a year. Display flight of ♂ accompanied by fluty trills. Nest: shallow, sparsely lined hollow in grass and heather. Eggs: 4, top-shaped, olive green with dark brown spots, incubated 26—28 days by both parents. Young led about 6 weeks. Diet: insects, worms, molluscs. Partial migrant. Migration: III—IV and VII—XI.

3 **WHIMBREL** *(Numenius phaeopus)* About size of pigeon. Resembles Curlew, but smaller, with quicker wing strokes, distinctly striped crown, whiter rump, darker tail and darker underside of wings. Downy plumage like Curlew's. Voice different, a rolling 'titti-titti-titti-tit' (best identification mark). Nests in Scandinavian tundra, keeps to coastal mud-banks during migration. Breeding season: V—VI, 1 brood a year. Nest and eggs like Curlew's. Eggs incubated (about 24 days) and young led (some 4 weeks) by both parents. Diet: worms, molluscs, crustaceans, insects. Migrant. Migration: IV—V and VII—IX.

1

2
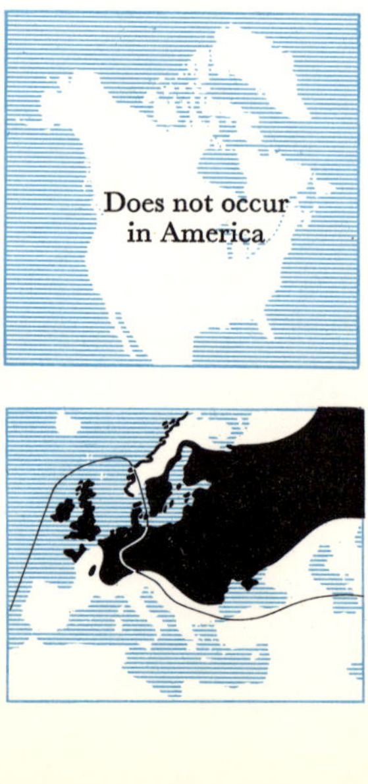

3
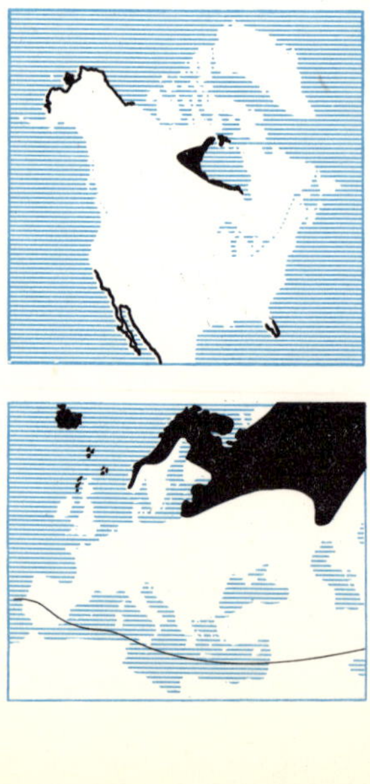

1
1
1 D
2
2 D
2
3
3

1 **BLACK-TAILED GODWIT** *(Limosa limosa)* Size of pigeon, with very long, straight beak and long legs. In nuptial plumage, neck and breast red. In winter plumage generally greyer. ♂ = ♀. Downy plumage rusty brown, with dark lores and spotted back. Distinguished in flight by broad white wing bands and white root and black end of tail. Voice: during display flight rhythmically repeated 'reeta'; warning calls hoarse 'kititit' and piercing 'quee-it'. Widespread summer bird in damp meadows, especially in lowlands, seen on mud of drained ponds during migration. Breeding season: IV—V, 1 brood a year. Nest: hollow in ground lined with dry grass. Eggs: 4, top-shaped, olive brown with dark brown spots, incubated about 24 days by both parents. Young led some 5 weeks. Diet: insects, worms, snails, slugs. Migrant. Migration: IV—V and VII—IX.

2 **BAR-TAILED GODWIT** *(Limosa lapponica)* Slightly smaller, with shorter beak and legs than Black-tailed Godwit. Body largely russet, wings and back greyish brown. Gently upcurved beak. Winter and juvenile plumage greyish brown. ♂ = ♀. Downy plumage like Black-tailed Godwit's. No wing stripe in flight, tail white with several bands. Voice (seldom heard): raucous 'kirrik' or 'yāk' on nesting site. Nests in Scandinavian wooded tundra, when migrating found mostly only on coast. Breeding season: V—VI, 1 brood a year. Nest and eggs like Black-tailed Godwit's. Eggs incubated (about 22 days) and young led by both parents. Diet: insects, crustaceans, worms. Migrant. Migration: IV—V and VII—X.

3 **SPOTTED REDSHANK** *(Tringa erythropus)* Rather larger than Blackbird, with thin, straight beak and long legs. Nuptial plumage black, with fine spots on back. Legs and base of beak red. Winter plumage slate grey on back, with white spots, underside white. Juvenile more spotted on underside than winter plumage. ♂ = ♀. Downy plumage like Redshank's. Relatively longer beak and legs than other sandpipers. Most easily identified by single-syllable call. Voice: energetic, fluty 'tchuitt'. Nests in Scandinavian forest marshes, regularly appears on muddy ponds and beaches when migrating. Breeding season: V—VI, 1 brood a year. Nest like Redshank's. Eggs: 4, olive brown, densely marked with dark brown spots, incubated by both parents. Diet: worms, molluscs, insects. Migrant. Migration: IV—V and VIII—X, often spends summer in western and central Europe.

1

2

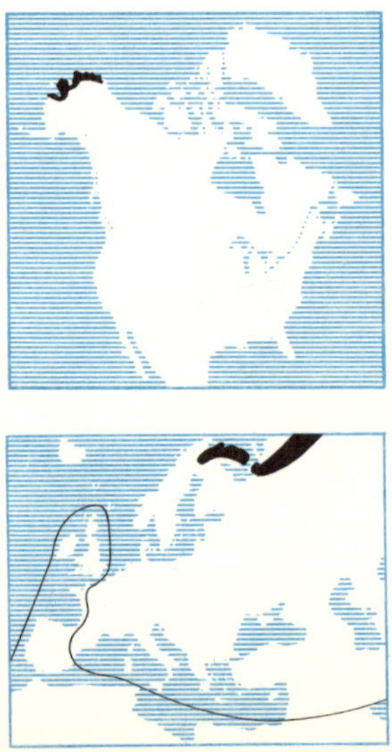

3

1 S
1 juv
1 D
2
2 juv
2 S
3 S
3 juv

1 **REDSHANK** *(Tringa totanus)* About size of Blackbird. Grey-brown, with dark spots and streaks; belly, rump and posterior border of wings white. Long legs and proximal half of beak red. Juvenile rust brown, with pale orange legs. Downy plumage yellowish brown, with black lines and reddish yellow legs. ♂ = ♀. In flight, long red legs and white-edged wings distinctive. Voice: melodious 'tleu-hu-hu', on nesting site long-repeated, warning 'teuk teuk teuk teuk'. Widespread summer bird in wet lowland meadows and by marshy ponds, seen when migrating on mud beside inland waters and on coast. Breeding season: IV—V, 1 brood a year. Nest: depression in grass, lined with a little straw. Eggs: 4, top-shaped, grey with dark spots, incubated 22—25 days by both parents. Diet: insects, worms, molluscs. Partial migrant. Migration: IV—V and VII—IX.

2 **MARSH SANDPIPER** *(Tringa stagnatilis)* Size of Starling, with straight, very slender, blackish beak and long grey legs extending well beyond light-coloured tail in flight. No border on wings. Winter and juvenile plumage generally lighter. ♂ = ♀. Downy plumage ochre brown and white, with dark stripes. Voice: clear 'tew-tew-tew', like call of Greenshank, but sweeter. Nests by lakes in east European steppes, seen occasionally in central and western Europe during migration. Breeding season: V—VI. Nest and eggs as Redshank. Both parents incubate eggs and tend young. Diet: aquatic insects, snails. Migrant. Migration: IV—V and VII—IX.

3 **GREENSHANK** *(Tringa nebularia)* Larger than Redshank, with long, slightly upcurved beak. Rather long, green legs. With pure white rump and underside appears light when flying. Head and breast whiter in winter. ♂ = ♀. Downy plumage like Redshank's, but lighter and with whiter head. Voice: loud 'tew-tew-tew'. Inhabits northern forest belt, nests in swamps and moors beside water, regularly seen in ponded regions and on coast when migrating. Breeding season: V—VI, 1 brood a year. Nest: on ground, thinly lined, hidden in low vegetation. Eggs: 4, top-shaped, yellowish grey with large, dark brown spots, incubated about 24 days by ♀. Both parents care for young. Diet: small aquatic and marsh organisms. Migrant. Migration: IV—V and VIII—X.

1

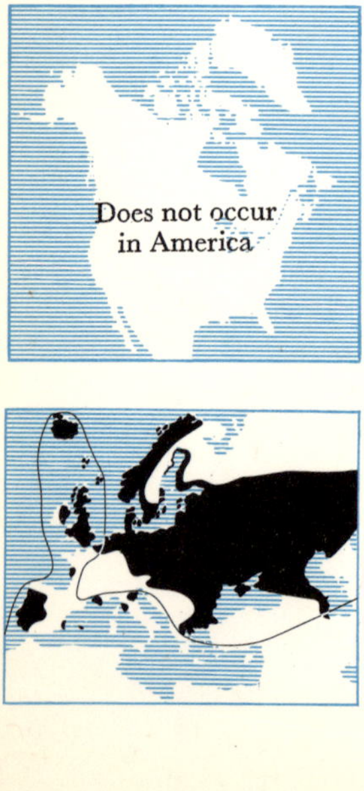

2

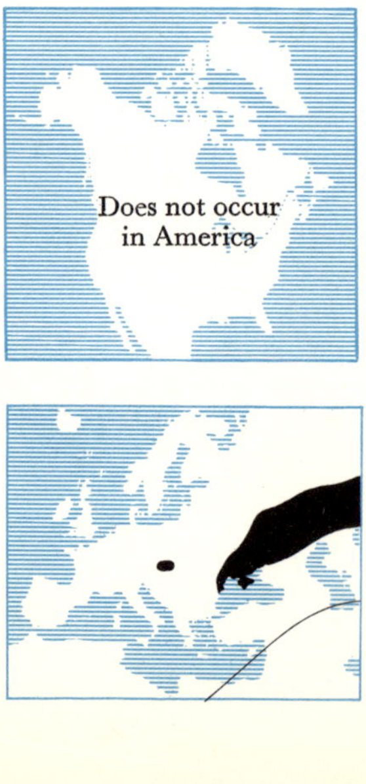

3

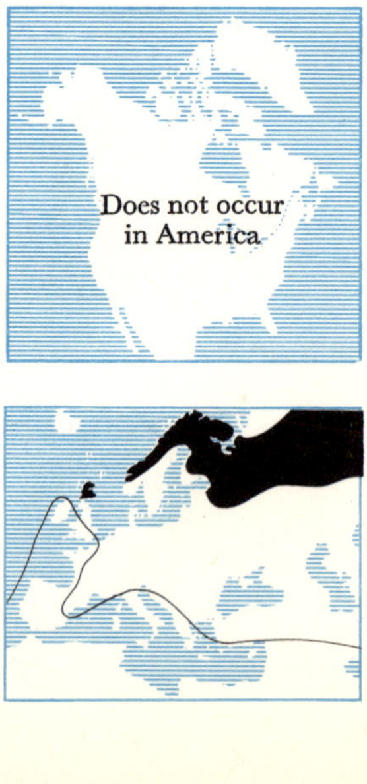

1 S
1 juv
1
1 D
2
2 S
2 juv
3 S
3 juv

Order: Waders - Charadriiformes **Family: Snipes and Sandpipers** - Scolopacidae

1 **GREEN SANDPIPER** *(Tringa ochropus)* Size of Song Thrush, with blackish brown, faintly spotted back. Underside and base of tail brilliant white. Beak and legs dark olive green. ♂ = ♀. Juvenile has distinct rust-coloured spots on back. Looks black from above when flying, underside of wings also dark and unstriped. Downy plumage light brown with black stripes. Voice: clear, whistling 'tlōōi-weet-weet'. Nests in marshy woods, moors and similar country near stretches of water. Seen on river banks and beside ponds during migration. Breeding season: IV—VI, 1 brood a year. Nest: rarely on ground, usually in tree, in other bird's (thrush's) nest. Eggs: 4, top-shaped, green with dark spots, incubated 21—24 days by both parents. Diet: aquatic insects, spiders. Partial migrant. Migration: IV—V and VIII—X.

2 **WOOD SANDPIPER** *(Tringa glareola)* About size of Crested Lark. Slimly built, with light eye stripe, brown, spotted back, white belly and rump. Legs yellowish grey, beak dark grey. Jerks body from side to side. Shows banded tail and light underside of unstriped wings in flight. ♂ = ♀. Juvenile has rust-edged dorsal feathers and distinctly striped neck. Downy plumage greyish white, with black stripes. Voice: clear 'chiff-iff-iff', display call (flying) 'tleea-tleea-tleea'. Nests in bogs and meadows, frequently seen on muddy ponds and coasts during migration. Breeding season: V, 1 brood a year. Nest: in shallow, sparsely lined hollow in ground. Eggs: 4, top-shaped, olive green with large brown spots. Incubation (21—24 days) and care of young by both parents. Diet: aquatic insects, spiders. Migrant. Migration: IV—V and VIII—X.

3 **COMMON SANDPIPER - SPOTTED SANDPIPER** *(Tringa hypoleucos)* Size of lark, with brown-grey back, grey breast and white belly. In flight, distinguished by white stripe, dark rump and white-edged tail. ♂ = ♀. Juvenile's dorsal feathers have narrow, rust-coloured border. Downy plumage yellowish grey, with dark crown and eye stripe. Most distinctive features are continual see-saw motions and water-skimming flight with jerky, shallow wingbeats. Voice: series of trills, 'titti-weeti' or 'twee-twee-twee', uttered mostly while flying and especially after dusk. Widespread nester on gravel-banks and sandy river and stream banks, seen beside lakes, ponds and rivers when migrating. Breeding season: V—VI, 1 brood a year. Nest: on ground, hidden. Eggs: 4, pear-shaped, brownish yellow with red-brown spots, incubated 20—23 days. Diet: worms, insects, molluscs. Migrant. Migration: IV—V and VII—X.

1

2

3

1
1 S
1 juv
2
2 juv
2 S
3
3 juv
3 D

1 **KNOT** *(Calidris canutus)* Size of Blackbird. In nuptial plumage has russet underside and brown, thickly spotted back (similarly coloured Curlew Sandpiper is smaller, with longer, curved beak and white rump). Winter plumage light, with white underside and grey, scale-marked back. ♂ = ♀. Recognizable in flight by stocky body and spotted rump. Juvenile has brown underside. Downy plumage mottled black and brown on back, crown and sides of head striped, underside white. Voice: seldom heard, sometimes calls 'twit-wit' when flying. Widespread coastal migrant, often in large, dense flocks, rarely seen inland. Nests in Arctic. Eggs: 4, grey-green with brown spots, incubated 20—25 days by both parents. Diet: chiefly small animals on shore. Migration: V and VIII—X.

2 **LITTLE STINT** *(Calidris minuta)* No larger than Sparrow, with rather short beak and short legs. In nuptial plumage russet brown on back and sides of head, belly white. Winter plumage more greyish brown. ♂ = ♀. Distinguished from Temminck's Stint by redder and more spotted back. Juvenile usually has conspicuous V-mark behind neck. Downy plumage like Dunlin's, but underside whiter. Voice: soft, rattling 'dirrdirrdirrit' when flying. Regular coastal bird of passage, also seen in small numbers inland, on mud beside water. Not timid as a rule. Nests in arctic tundra. Breeding season: VI—VII. Nest: on ground, in low vegetation near water. Eggs: 4, brownish green with brown spots. Both parents tend brood. Incubation and rearing period not known. Diet: mostly small shore animals, a few seeds. Migrant. Migration: V—VI (in small numbers) and VII—X.

3 **TEMMINCK'S STINT** *(Calidris temminckii)* About size of Sparrow, resembles Little Stint, but always much greyer, with less vividly spotted back and greyer head and sides of throat. Shows white tail feathers and indistinct white wing stripe when flying. Beak dark brown, legs greenish brown. Voice: frequently repeated 'tirrr' reminiscent of grasshopper's chirping. Not a gregarious bird. Nests in Scandinavian tundra, regularly visits coasts during migration and sometimes seen inland on mud-banks in lakes and ponds. Breeding season: VI—VII. Nest: on ground, hidden in low vegetation. Eggs: 4, green with small brown spots, mostly at blunt end. Diet: small worms, molluscs and insects. Migrant. Migration: IV—V and VII—IX.

1

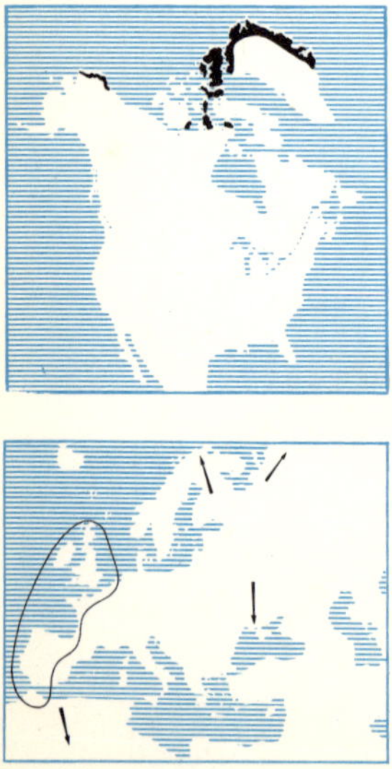

2

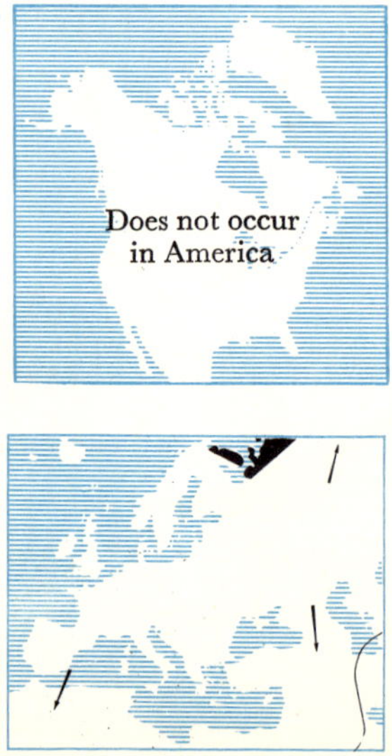

3

1 S
1 juv
1 S
2 S
2 S
2 juv
3 S
3 S
3 juv

1 **DUNLIN** *(Calidris alpina)* About size of Starling, with gently down-curved beak. In nuptial plumage has rust brown back with blackish brown spots and black belly. Legs and beak black. Winter plumage grey-brown. ♂ = ♀. Juvenile has rust-coloured, white-striped back. Downy plumage light brown with dark stripes. Voice: frequent soft 'treer', display call purring 'twirr-wirr-wirr'. Nests in Scandinavian tundra, near coast, frequent bird of passage on coasts, also inland, on drained ponds; often occurs in large flocks. Breeding season: IV—VI, 1 brood a year. Nest: small, straw-lined depression in grass. Eggs: 4, olive brown with large, dark spots, incubated 17—20 days by both parents. Diet: small animals on shore and in mud, seeds. Partial migrant. Migration: VII—XI and IV—V.

2 **CURLEW SANDPIPER** *(Calidris ferruginea)* Size of Starling. Distinguished in nuptial plumage by russet underside and down-curved beak. In winter and juvenile plumage might be mistaken for Dunlin, but has longer legs, unstriped, white rump and less spotted, more rusty yellow breast. ♂ = ♀. Downy plumage rust brown, with wide, dark stripes and white stippling. Voice: metallic 'chirririp' or 'chirrip'. Nests in tundra belt of northern Asia; regular, though rare, migrant to central and western Europe. Frequents flat shores and muddy, drained ponds. Breeding season: VI—VII. Nest: in moss, between clumps of grass. Eggs: 4, olive green with black-brown spots. ♀ incubates eggs and leads young. Diet: small animals on shore and in mud, insects, seeds. Migrant. Migration: IV—V and VII—IX.

3 **SANDERLING** *(Calidris alba)* Size of Starling. Very light, with pure white belly and short, shiny black legs and beak. Nuptial plumage reddish brown on back and breast, with black spots and streaks. Winter plumage light grey on back, with black spot on carpal joint. Juvenile has black-spotted back. Conspicuous wing stripe and dark, greyish white-sided tail seen during flight. Downy plumage light brown, with black-spotted back. Voice: sharp 'twick'. Nests in arctic coastal regions, less often inland. Breeding season: VI—VII. Nest: shallow depression in moss, among grass, lined with dry vegetation. Eggs: 4, olive green with a few brown spots, incubated 23—24 days. Young reared by both parents. Diet: insects, molluscs, crustaceans, sometimes parts of plants and seeds. Migrant. Migration: III—V and VIII—X.

1

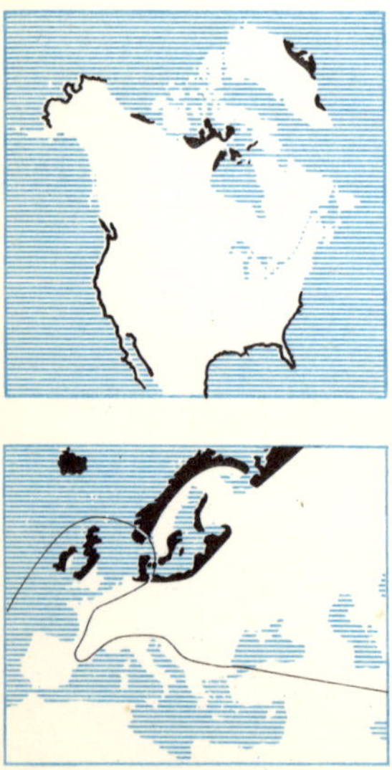

2

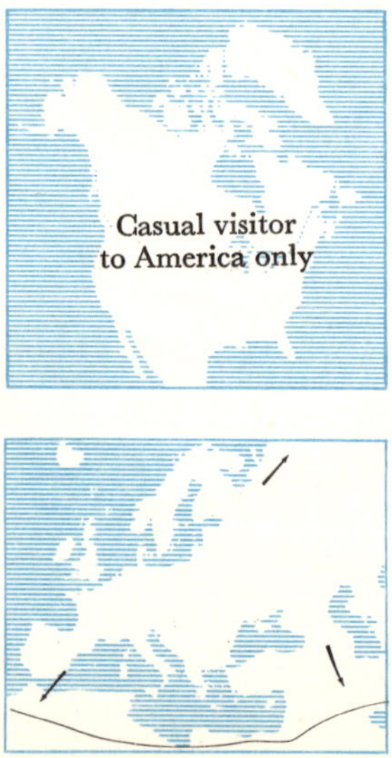

3

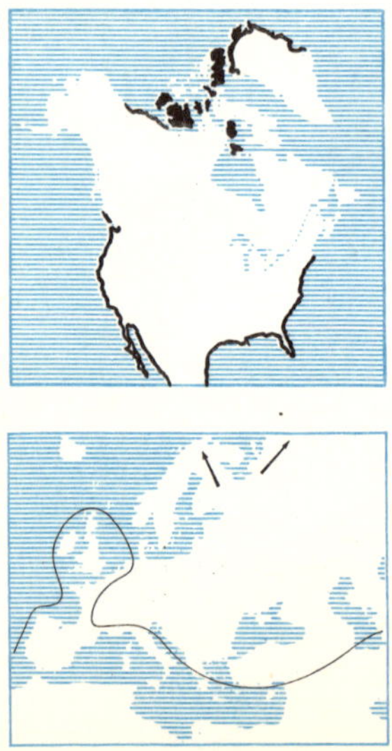

1 S
1 juv
2 W
2 juv
2 S
3 S
3 W
3 S
3 juv

Order: Waders - Charadriiformes **Family: Snipes and Sandpipers** - Scolopacidae

1 **RUFF** *(Philomachus pugnax)* Size of Blackbird. Fairly long legs, straight beak same length as head. ♂ in nuptial plumage has bright, expansible ruff and eartufts (looks thick-necked in flight). ♀ and winter plumage brown-backed, with dark spots. ♀ smaller than ♂. Juvenile fawn-backed, with rusty yellow breast and white belly. Downy plumage yellowish brown on back, with dark blotches, underside rusty yellow. Voice: a rare, soft 'kuk-uk-uk', but mostly silent. Widespread nester in lowland meadows and swamps, often seen inland on ponds when migrating. Breeding season: V—VI, 1 brood a year. Nest: on ground, poorly lined. Eggs: 4, top-shaped, grey-green to olive brown, spotted, incubated 21 days by ♀. Diet: insects, worms, sometimes seeds. Migrant. Migration: III—IV and VIII—X.

Order: Waders - Charadriiformes **Family: Avocets and Stilts** - Recurvirostridae

2 **AVOCET** *(Recurvirostra avosetta)* About size of domestic pigeon, with long neck, long blue-grey legs and thin, up-curved beak. Nuptial plumage pure white, with sharply demarcated coal-black feathers on head, shoulders and wings. Fluttering flight, like Lapwing. ♂ = ♀. In winter and juvenile plumage black areas more blackish brown. Downy plumage sandy, with fine black lines, underside white. Voice: tuneful 'kloo-it', on nesting site warning 'kleep-kleep-kleep'. Nests in colonies in coastal meadows, rarely inland by salt lakes. Breeding season: IV—VI, 1 brood a year. Nest: shallow depression in sand, lined with straw. Eggs: 3—4, clay-brown with a few dark spots, incubated 22—24 days by both parents. Diet: insects, small crustaceans and molluscs, gathered in shallow water by moving head from side to side. Partial migrant. Migration: IV and VII—X.

3 **BLACK-WINGED STILT - BLACK-NECKED STILT** *(Himantopus himantopus)* About size of pigeon. Very long, stilt-like, red legs and thin, straight beak. Plumage black and white, little difference between ♂ = ♀. In juvenile, black markings more dark brown. Downy plumage sandy, with small spots, underside white. Walks with dignified strides, in flight legs trail behind tail. Voice: repeated, clear 'kyip'. Irregular summer bird in flood areas and by salt lakes, nests in colonies. Seldom seen during migration. Breeding season: V, 1 brood a year. Nest: shallow, straw-lined depression. Eggs: 3—5, greenish yellow with dark brown spots. Eggs incubated (25—26 days) and young led (almost 1 month) by both parents. Diet: mainly insects, also small molluscs and worms. Migrant. Migration: IV—V and VII—IX.

1

2

3

1 ♂ S
1 ♂ S
1 ♂ S
1 ♂ W
1 ♂ S
1 D
1 ♀
2
2 D
3
2 juv

1 **GREY PHALAROPE - RED PHALAROPE** *(Phalaropus fulicarius)* Barely size of thrush, with large yellow beak. Nuptial plumage: white cheek spots extending up over eyes, whole of underside dark russet. ♂ distinguishable by somewhat smaller size and duller colouring. Winter plumage dull grey on back; head (except for dark eye stripe) and underside white. Juvenile has more or less brown back. Downy plumage rusty yellow with dark spots and stripes, underside greyish white. Voice: piercing 'whit', uttered serially on nesting site. Nests in bog meadows in tundra, seen on coasts, seldom inland when migrating. Breeding season: VI—VII. Nest: sparsely lined hollow in ground. Eggs: 4, brown, speckled, incubated about 19 days by ♀. Diet: small crustaceans and molluscs, worms, insects. Migrant. Migration: III—VI and VIII—XI.

2 **RED-NECKED PHALAROPE - NORTHERN PHALAROPE** *(Phalaropus lobatus)* Size of lark, with very thin, black beak. Nuptial plumage: underside white, with russet neck-band, white cheek spot does not extend to eyes. Colouring of ♂ duller, neck-band more greyish brown. In winter and juvenile plumage closely resembles Grey Phalarope, but back blacker and eye stripe more sharply defined. Downy plumage like Grey Phalarope's. Voice: short, monosyllabic 'tchwick'. Same habitats as Grey Phalarope, but seen more frequently when migrating. Breeding season: V—VI. Nest: in clump of grass, usually near water, lined with straw. Eggs: 4, stone grey to olive brown with dark brown spots. Care of brood as Grey Phalarope. Diet: chiefly insects and small crustaceans. Migrant. Migration: V and VII—XI.

Order: Waders - Charadriiformes **Family: Pratincoles** - Glareolidae

3 **PRATINCOLE** *(Glareola pratincola)* About size of thrush, with large head, short beak, long, pointed wings and forked tail. Resembles large swallow in flight. Grey-brown, with white rump and belly. Wings almost black, under tail-coverts russet, throat creamy yellow outlined with black. Juvenile has black-speckled breast and back. Downy plumage sandy brown, with dark stripe over head and down back. Voice: noisy 'kitti-kirrik'. Gregarious bird, nests in dry salt-steppes, accidental bird in central and western Europe. Breeding season: V—VI, 1 brood a year. Nest: on bare ground. Eggs: 2—3, greyish yellow, thickly speckled. Eggs incubated (about 18 days) and young tended (some 25 days) by both parents. Diet: insects caught on wing or on ground (e.g. grasshoppers). Partial migrant. Migration: IV—V and IX—X.

1

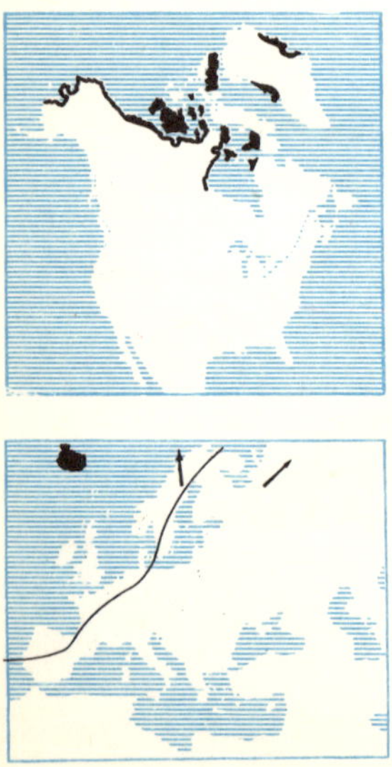

2

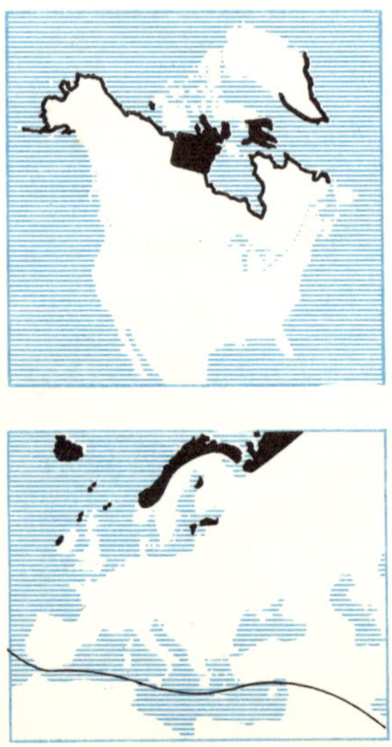

3

1 W
1 ♀ S
2 W
2 W
2 ♀ S
3
3 juv
3

Order: Waders - Charadriiformes

Family: Stone Curlews - Burhinidae

1 **STONE CURLEW** *(Burhinus oedicnemus)* Size of Wood Pigeon. Ungainly, with short, thick beak and thick legs (especially tarsal joint). Wide head with large, yellow eyes. Sandy brown, dark-streaked. Shows 2 white wing bands in flight. ♂ = ♀. Juvenile more fawn, less spotted. Downy plumage sandy, with 2 narrow, black dorsal stripes. Secretive, likes twilight. Voice: harsh, loud 'coo-ree', heard mainly at night. Rare nester on barren ground with little vegetation and on heaths; not bound to water. Breeding season: IV—VII, often 2 broods a year. Nest: shallow hollow sparsely lined with small stones and fragments of plants. Eggs: 2, light grey with spots and scribbles. Eggs incubated (25—27 days) and young led (about 1 month) by both parents. Diet: insects, snails, small vertebrates. Partial migrant. Migration: IV and VII—X.

Order: Gulls and Terns - Lariformes

Family: Skuas - Stercoraridae

2 **ARCTIC SKUA - PARASITIC JAEGER** *(Stercorarius parasiticus)* Size of Black-headed Gull. Mostly dark brown, resembles gull. Middle tail feathers in adults several cm longer than others, form point. Shows white patch on wings when flying. Light variant has white underside and black crown. Juvenile has thickly banded underside; middle pair of tail feathers only just discernibly longer. Downy plumage dark brown. Voice: cat-like 'ya-wow', also 'gack-gack'. Nests in Scandinavian tundra, seen on sea and coast when migrating (rarely inland). Breeding season: V—VI. Nest: shallow depression in moss or grass. Eggs: 2, brownish green with dark spots. Eggs incubated (24 days) and young tended by both parents. Diet: chiefly fish snatched from other sea-birds. Migration: mainly VIII—X.

3 **POMARINE SKUA - POMARINE JAEGER** *(Stercorarius pomarinus)* Size of Common Gull. Closely resembles Arctic Skua, but middle pair of tail feathers, in adults, stretches far beyond end of tail and is blunt-ended and twisted. Juvenile hard to distinguish from Arctic Skua. Downy plumage light brown and grey-brown. Voice: gull-like 'yew', on nesting site warning 'week-week-week'. Nests in far northern tundra, seen as migrant on coasts, seldom on inland water. Breeding season: VI—VII. Nest, eggs, care of brood and diet as Arctic Skua. Migration: mainly in IX and X.

1

2

3

1
1 D
2 DV
2 imm
2 LV
2 DV
3 LV
3 DV
3 LV
3 imm

Order: Gulls and Terns - Lariformes **Family: Skuas -** Stercoraridae

1 **LONG-TAILED SKUA - LONG-TAILED JAEGER** *(Stercorarius longicaudus)* Size of Black-headed Gull. Adults: pointed middle tail feathers project beyond rest for more than double tail length. Black cap, white neck, light underside. Juvenile and downy plumage like Arctic Skua. Voice: sometimes 'kri-kri-kri' or 'kr-r-r-r', from nesting site shrill 'kree'. Nests in Scandinavian tundra, when migrating occasionally seen on coast and open sea, seldom inland. Breeding season: VI. Nest and eggs like other skuas'. Eggs incubated about 23 days by both parents, care of young about 3 weeks. Diet: small vertebrates; insects, worms; otherwise parasitic. Migration: V and VIII—XI.

2 **GREAT SKUA - SKUA** *(Stercorarius skua)* Size of large Herring Gull. Robust, dark brown, with white feathers at base of primaries. Fairly wide, blunt-tipped wings. Middle tail feathers not discernibly lengthened. Thick beak. Juvenile normally indistinguishable. Downy plumage yellowish brown. Voice: raucous 'hah-hah-hah' and 'skerr'. Nests on North Sea islands; when migrating seen on sea, occasionally on coast, seldom strays inland. Breeding season: V—VI. Nest and eggs like other skuas'. Eggs incubated up to 30 days and young tended 6—7 weeks by both parents. Diet: mostly fish (snatched from other sea-birds), birds, birds' eggs and young. Migration: VIII—IV.

Order: Gulls and Terns - Lariformes **Family: Gulls -** Laridae

3 **GLAUCOUS GULL** *(Larus hyperboreus)* About size of Great Black-backed Gull. Very pale, none of plumages has black-tipped wings or black-edged tail. Yellow beak. Downy plumage like Great Black-backed Gull's, but lighter. Voice similar to Herring Gull's, shrill 'gak-gak-gak' and wailing 'kee-ow'. Nests on Scandinavian coasts and rocky islands; seen when migrating on open sea, less on coast, inland as accidental bird. Breeding season: V—VII, nests in colonies. Nest: pile of moss, seaweed and turf. Eggs: 3, greenish to brown, with dark spots, incubated 28 days (and young fed) by both parents. Diet: fish, birds, eggs, carrion, also insects and berries. Migration: X—III.

1

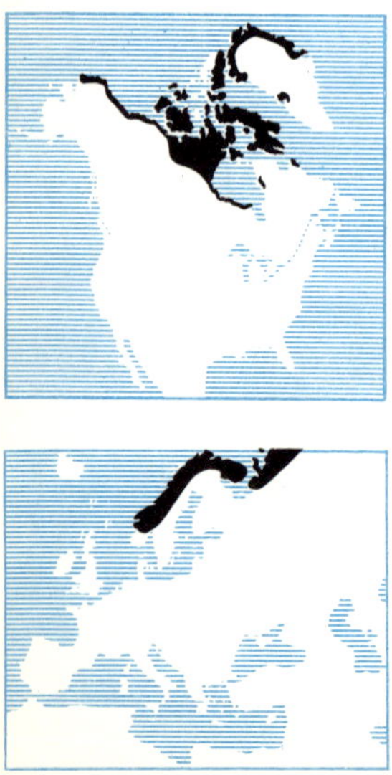

2

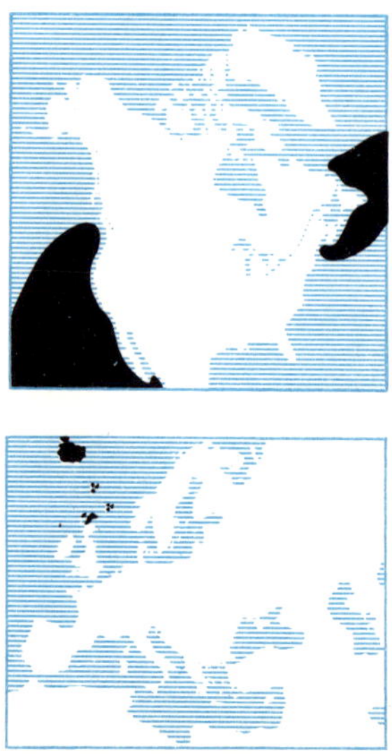

3

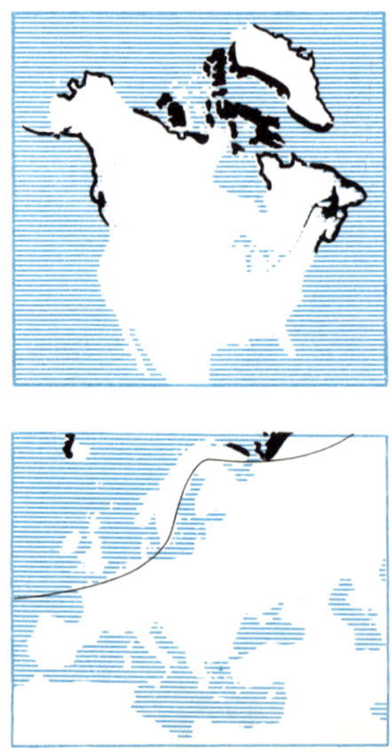

1
2
1 juv
3 S
3 juv
3 two years old

1 **GREAT BLACK-BACKED GULL** *(Larus marinus)* Larger than Herring Gull, with much larger, yellow beak. Black mantle. Legs flesh pink. Juvenile distinguished from Herring Gull by size and whiter underside. Downy plumage similar to Herring Gull's, but greyer and with smaller head spots. Voice: deep, chattering and mournful, serial 'gak-owk-ow-kyow-yowk'. Common sea-bird, nests on rocky coastal islands, keeps to coast when migrating, rarely appears inland. Breeding season: V—VI, 1 brood a year. Nest: pile of plant fragments, with a few feathers. Eggs: 3, brown with dark spots, incubated 26—28 days by both parents. Young fed about 50 days. Diet: various vertebrates, worms, molluscs, refuse. Partial migrant. Migration: VII—IV.

2 **LESSER BLACK-BACKED GULL** *(Larus fuscus)* Size of Herring Gull. Slate grey to black mantle, yellow beak, yellow legs. Smaller and more active than Great Black-backed Gull, with thinner beak. Juvenile brown, with black beak and dull pink legs. Hard to distinguish from young Herring Gull, but has blacker primaries. Downy plumage like Herring Gull's. Voice: like Herring Gull's, but slightly higher and thinner. Nests in colonies on sea-coast, less often by large inland seas and on moors; regularly seen on inland waters during migration. Breeding season: V—VI. Nest, eggs and care of young as Herring Gull. Migrant. Migration: IV—V and IX—X.

3 **HERRING GULL** *(Larus argentatus)* About size of buzzard, with light grey mantle, large yellow beak and flesh pink or yellow legs. In winter plumage, like all large gulls, has grey-brown streaks on head and neck. Juvenile plumage brown, with dark spots and blackish brown primaries and tail band; worn at least 2 years, lighter with each moult (transitional plumage). Downy plumage cream, with dark spots on head and back. Voice: loud, serial 'kyow-kyow-gah-gah-gah'. Commonest seagull. Sometimes nests in large colonies on coast, seen in mouths of large rivers and occasionally inland when migrating. Breeding season: V—VI, 1 brood a year. Nest: pile of plant material. Eggs: 3, olive green to brown, with blackish brown spots, incubated 25—27 days by both parents. Care of young 6 weeks. Diet: various marine creatures, birds' eggs and young. Partial migrant. Migration: III—IV and VIII—X.

1

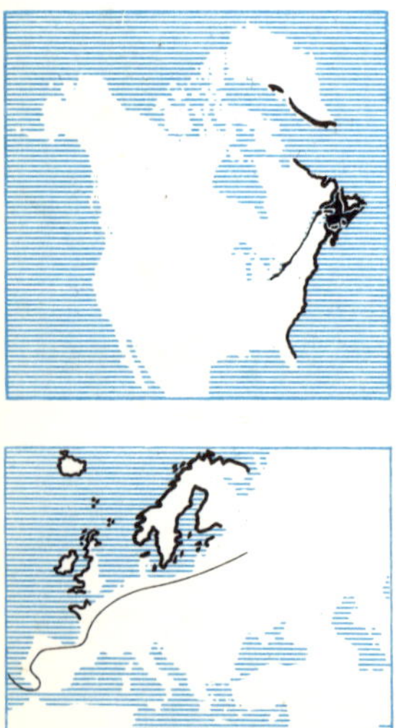

2

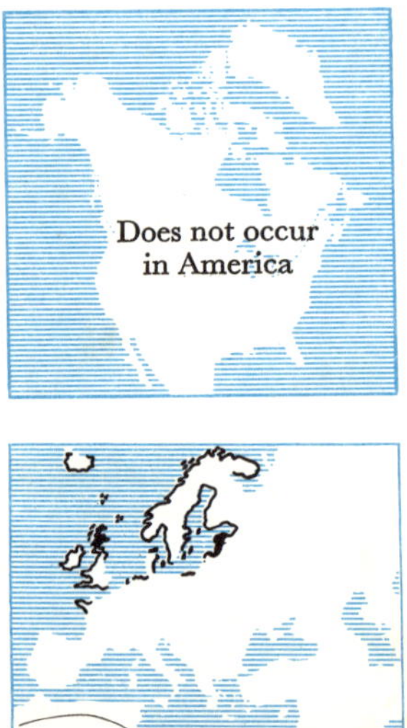

3

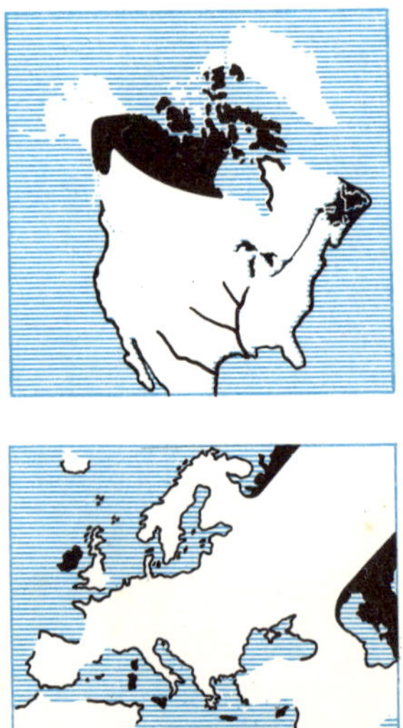

1 S
1 imm
1
2 imm
2 S
2
3 imm
3
3 S
3 D

1 **COMMON GULL - MEW GULL** *(Larus canus)* Smaller than Herring Gull, larger than Black-headed Gull. Adults have grey mantle and long, black tips on wings. Legs and thinnish beak greenish grey. Winter plumage has dark streaks on head and neck. Juvenile mainly brown, dark-spotted, with black band at end of tail; beak bluish with black-brown tip, legs flesh pink. Downy plumage like Herring Gull's. Voice: 'keee-ya, keeow', higher than Herring Gull's. Common nester on coast and a few inland seas, often in large colonies; regularly winters on inland waters, also occurs in towns. Breeding season: V—VI, 1 brood a year. Nest, eggs and care of young as Herring Gull; eggs incubated 22—23 days, young fed about 5 weeks. Diet: like Herring Gull's, more insects. Partial migrant. Migration: III—IV and VII—XI.

2 **BLACK-HEADED GULL** *(Larus ridibundus)* Size of pigeon. In summer plumage has blue-grey mantle and coffee-coloured cap. Beak and legs crimson. Winter plumage has white head and dark ear spot. Shows white anterior border of wings in flight. Juvenile more or less brown on head, back and upper surface of wings; black band at end of tail. Downy plumage rust brown, with dark-spotted back; beak and legs dull pink (as distinct from Common Tern). Voice: hoarse 'krree-ah, kwarr, kek-kek', calls incessantly. Nests in colonies by overgrown, marshy ponds and lakes; seen on rivers, in fields and in towns when migrating. Breeding season: IV—VI, 1 brood a year. Nest: pile of plant material, often on sedge. Eggs: 3, usually olive green with brown spots, incubated 23 days by both parents. Young tended up to 6 weeks. Diet: insects, worms, young fish. Partial migrant. Migration: III and VII—X.

3 **MEDITERRANEAN GULL** *(Larus melanocephalus)* About size of Black-headed Gull, but more robust, with thicker, coral beak. Coal black cap covers back of head. Legs dark red. Winter and juvenile plumage similar to Black-headed Gull's. No white border or black tips seen on wings during flight. Downy plumage greyish yellow, with few dark markings. Rare nester in marshes and lagoons near Mediterranean coast, occasionally nests inland in Black-headed Gull colonies; seen inland as accidental migrant. Breeding season: V—VI. Nest and care of brood as Black-headed Gull. Eggs more sandy, with small spots and streaks. Diet: small fish, molluscs, insects. Partial migrant. Migration: IX—IV.

1

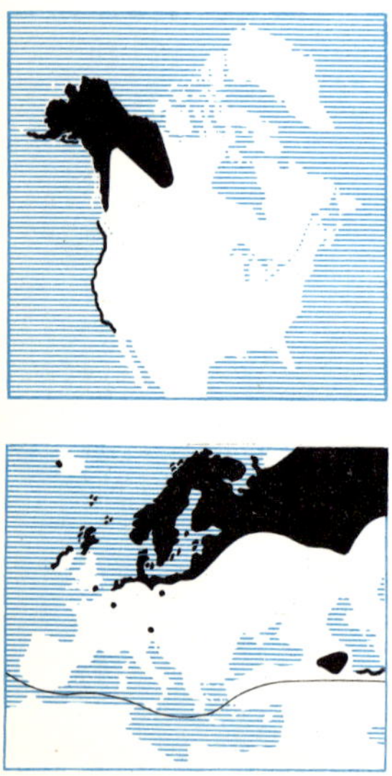

2

3

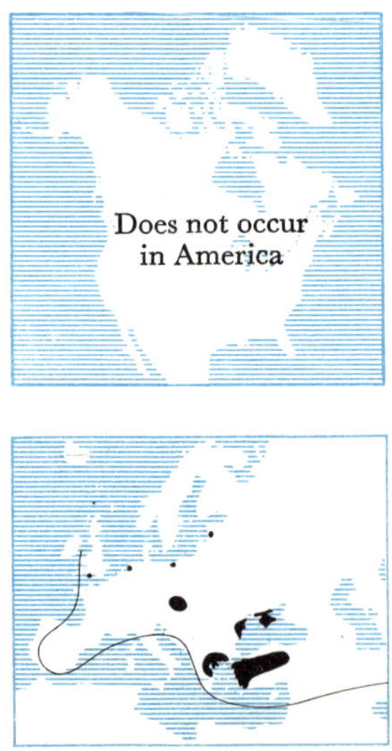

1 imm
1 D
1 S
2 W
2 imm
2 D
2 S
3 imm
3 W
3 S

1 **LITTLE GULL** *(Larus minutus)* Smaller than Black-headed Gull. Black cap reaches to nape, beak reddish brown, feet vermilion. Winter plumage: head white, with grey crown and ear spot. Juvenile has black-brown back and black band at end of tail. Underside of wings dark in flight, wing strokes quicker than Black-headed Gull; juvenile has blackish M-stripe on wings. Downy plumage light brown with dark spots, legs flesh pink. Voice: an occasional soft 'kek-kek-kek, kā-kā-kā'. Nests in small, isolated colonies on thickly overgrown lakes, ponds and moors; when migrating, seen in small numbers on lakes and ponds. Breeding season: V—VI, 1 brood a year. Nest: pile of blades of marsh plants. Eggs: 3, olive green, spotted, incubated 21—23 days by both parents. Young tended up to 25 days. Diet: chiefly insects. Partial migrant. Migration: IV and VIII—XI.

2 **KITTIWAKE - BLACK-LEGGED KITTIWAKE** *(Rissa tridactyla)* Larger than Black-headed Gull. Mantle darker than Common Gull's, beak waxy yellow, legs black. Black-tipped wings with no white spots. Winter plumage: head has grey and blackish spots in front of and behind eyes, light grey nape. Juvenile: black nuchal band, wing stripe and terminal tail band, black beak. Downy plumage grey-brown, unspotted on back, underside white. Voice: on nesting site distinctive 'kitti-wāk', otherwise mostly silent. Common nester on coasts of north-western Europe, keeps to open sea during migration; sometimes driven inland by storms. Nest: pile of plant fragments on cliff. Eggs: 2, greyish white with round spots. Eggs incubated (21—24 days) and young tended (4—5 weeks) by both parents. Diet: small fish, crustaceans, etc. Partial migrant. Migration: VIII—VI.

Order: Gulls and Terns - Lariformes **Family: Terns** - Sternidae

3 **CASPIAN TERN** *(Hydroprogne tschegrava)* Size of Common Gull, largest tern. Large red beak, dark brown legs, slightly forked tail. Primaries look dark on underside. Winter plumage: deep black of cap intermingled with white. ♂ = ♀. Juvenile: cap and dorsal plumage intermingled with dark brown feathers, beak dull orange. Downy plumage grey to white, beak bright red. Voice: loud, raucous 'kraa-uh, klee-eep', etc. Nests in colonies on sandy shores and islands; a few birds visit inland waters during migration. Breeding season: V—VI, 1 brood a year. Nest: shallow depression in sand. Eggs: 2—3, yellowish grey with dark brown spots, incubated 22—24 days by both parents. Diet: fish, caught by hovering, sometimes young sea-birds. Migrant. Migration: IV—V and VIII—X.

1

2

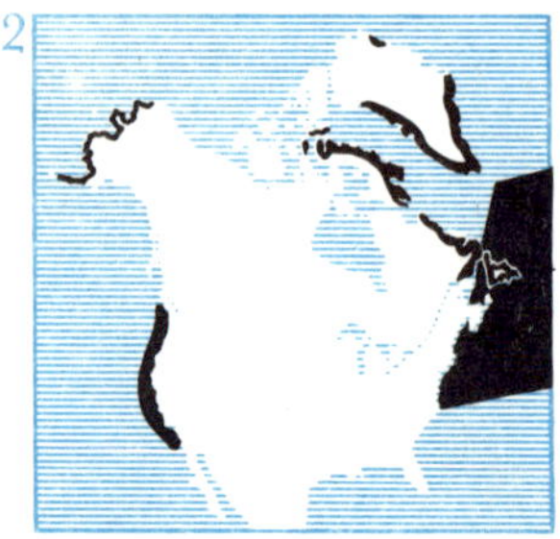

3

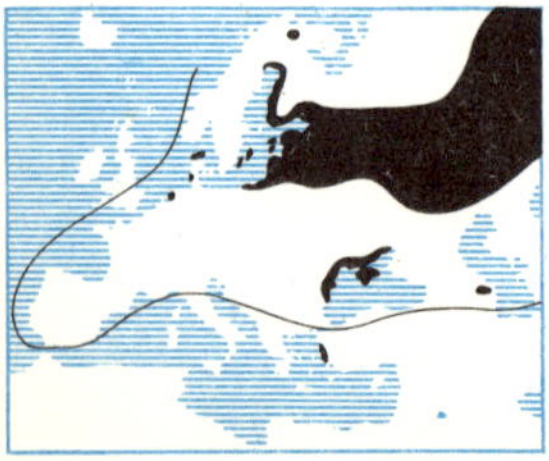

1 S
1 S
1 transitional plumage
1 imm
2 S
2 S
2 imm
3 imm
3 S

1 **WHITE-WINGED BLACK TERN** *(Chlidonias leucopterus)* Size and appearance like Black Tern, but tail and upper wing-coverts pure white. Body and under wing-coverts coal black. In winter plumage no black spots on sides of crop. Juvenile and downy plumage closely resemble Black Tern's. Voice: snarling 'kik kik' and 'kerr'. Sporadic nester on marshy lakes and ponds, also much rarer than Black Tern during migration. Breeding season: V—VI. Nest, eggs, care of young and diet as Black Tern. Migrant. Migration: V—VI and VIII—IX.

2 **WHISKERED TERN** *(Chlidonias hybrida)* Similar to Black Tern in size and appearance, but much paler than White-winged and Black Tern. Black-capped head; white cheeks and sides of neck form broad white band, underside of wings also white, tail grey. Winter plumage: no dark spots on sides of crop, little black on crown, grey nape and plain, light back. Juvenile: black spots on back and upper surface of wings. Downy plumage rusty yellow, spotted, throat black, underside pure white. Voice: harsh 'schreea' and 'ky-ik'. Rare nester on marshy rivers and ponds. Breeding season: V—VI. Nest: pile of material on floating plants. Eggs: 3, bluish green, spotted, incubated at least 18 days (and young fed) by both parents. Diet: chiefly insects. Migrant. Migration: IV—V and VIII—IX.

3 **BLACK TERN** *(Chlidonias niger)* Much smaller and daintier than Black-headed Gull. Body black, upper surface of wings and tail slate grey, underside of wings grey, under tail-coverts white. ♀ slightly greyer than ♂. Winter plumage: back grey, head and nape speckled, underside white. Juvenile: brown-speckled crown and back, beak rusty brown. Downy plumage ochre yellow with black spots. Voice: 'kirr', or 'kik'. Not common, nests in small colonies on overgrown ponds and lakes. Breeding season: V—VI, 1 brood a year. Nest: on floating plants. Eggs: 3, olive brown, with dense dark brown spots, incubated 14—17 days by both parents. Young fed about 1 month. Diet: mainly insects caught on surface of water in short, regular swoops, also fish fry and small fish. Migrant. Migration: V and VII—IX.

1
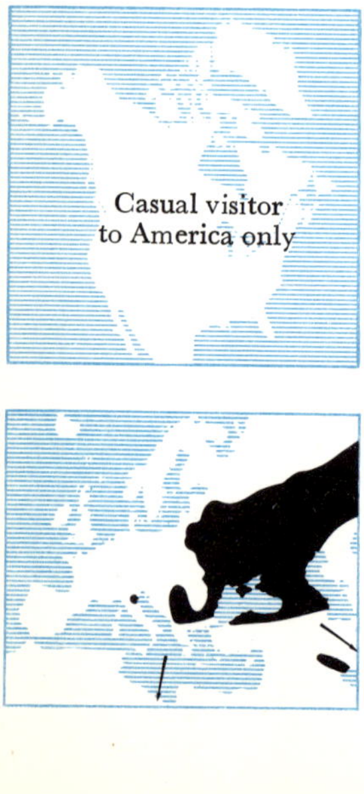

2

3
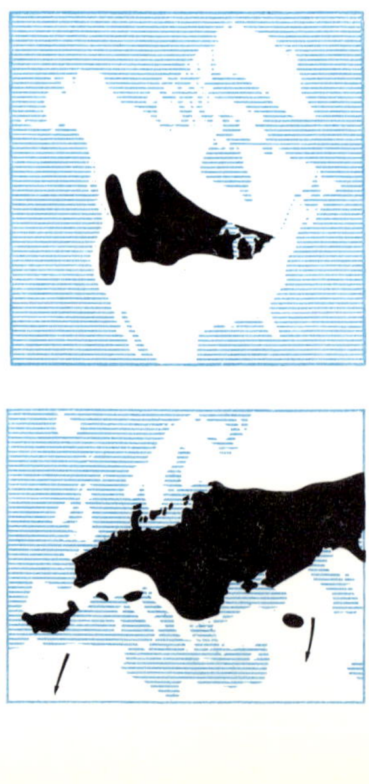

1 S
1 S
1 imm
2 S
2 S
2 imm
3 S
3 S
3 W
3 imm
3 D

1 **GULL-BILLED TERN** *(Gelochelidon nilotica)* Size of Black-headed Gull, with black, rather high-ridged beak and slightly forked, light grey tail. Legs black, relatively long. Sitting bird's wing tips project far beyond short end of tail. Resembles gull in flight. Winter plumage: white forehead, whitish grey back. Juvenile: mostly whiter, head sometimes cream, back dark-spotted. Downy plumage yellowish brown, mostly dark-spotted, underside almost white. Voice: loud, laughing 'ka-huk, ka-huk'. Nests in colonies on sandy shores and islands, on lakes in steppes, occasionally on river islands. Breeding season: V—VI, 1 brood a year. Nest: hollow in sand or short grass. Eggs: 3, sandy with brown spots, incubated 22—23 days (and young fed about 5 weeks) by both parents. Diet: mainly insects, also small vertebrates, etc. Migrant. Migration: V and VII—IX.

2 **COMMON TERN** *(Sterna hirundo)* Smaller than Black-headed Gull, slim-bodied, with narrow wings and light, buoyant flight. White, deeply forked tail, beak brick red, with black tip. Winter plumage: white forehead, black nape, proximal end of upper forewing noticeably darker than rest; beak brown to black. Juvenile distinguished by brown feathers on back and crown. Downy plumage fawn, with black spots on back; beak and legs salmon pink. Voice: loud, screeching 'kit-kit-keerr, kik' or 'kēarrr'. Widespread nester on coasts and many river islands, ponds and lakes, seen mostly on coast when migrating. Breeding season: V—VII, 1 brood a year. Nest: shallow, poorly lined depression in ground. Eggs: 3, light to olive green with dark spots, incubated 20—22 days (and young fed 1 month) by both parents. Diet: small fish, crustaceans, insects. Migrant. Migration: IV—V and VII—IX.

3 **ARCTIC TERN** *(Sterna paradisaea)* Very similar in size and appearance to Common Tern. Differences subtle, e.g. sides of head along edge of black cap often distinctly whiter than rest of underside, beak finer and usually all red. Legs extremely short (hardly visible when standing), tail tip longer than wing tips. Winter and juvenile plumage like Common Tern's. Downy plumage: minute dark spots on back. Voice: softer than Common Tern's, short 'kik-kik-kirrā'. Nests in large colonies on coasts, occurs inland as accidental bird only. Breeding season: V. Nest, eggs and care of young as Common Tern, usually only 2 eggs. Diet: small fish, crustaceans, a few insects. Migrant. Migration: IV—V and VII—X.

1

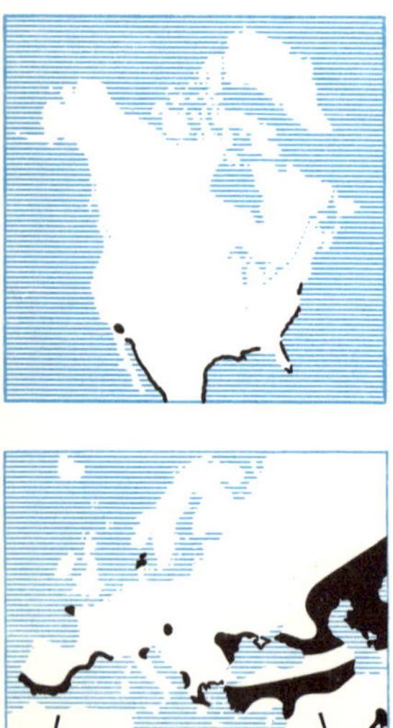

2

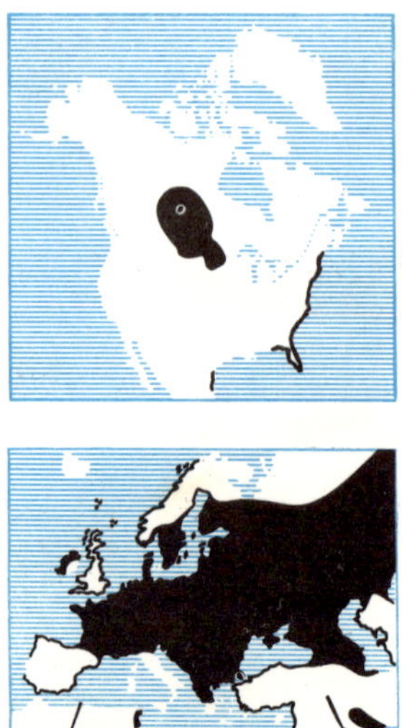

3

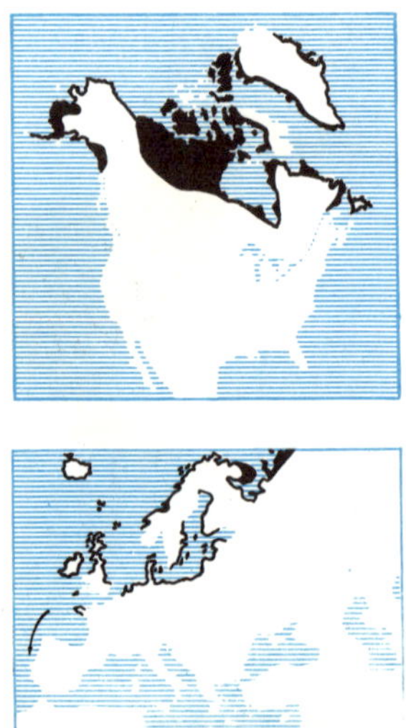

1 S
1 W
1 imm
2 W
2
2 S
2 imm
2 D
3 S
3
3 imm

1 **ROSEATE TERN** *(Sterna dougallii)* In all plumages, size and form closely resembles Common Tern, but has black, thinner beak red only at base. Very long outer tail feathers project beyond wings in sitting bird. Underside of nuptial plumage often pinkish. Voice: screeching 'kreee', also soft 'chū-ick' (quite unlike calls of similar Common and Arctic Tern). Rare and irregular nester on European coasts. Breeding season: VI—VII. No nest, 2 cream, russet-spotted eggs laid on sand, incubated 21 days by both parents. Diet: small sea-fish. Migrant. Migration: IV—V and VIII—IX.

2 **LITTLE TERN - LEAST TERN** *(Sterna albifrons)* Size of Swift, smallest tern. Nuptial plumage distinguished by white, sharply but unevenly contoured forehead, thin, waxy yellow, black-tipped beak and yellow legs. Forked tail same length as wings in sitting bird. Winter plumage: ash grey crown, black nape. Juvenile has black and brown spots on back and wings. Downy plumage sandy, underside white, fine dark markings on back. Voice: high, rattling 'kirri-kirri-kirri' or single 'kree-ik' calls, etc. Breeding season: V—VI, 1 brood a year. Nest: shallow depression in sand, lined with a few plant and shell fragments. Eggs: 2—3, sandy, dark-spotted, incubated 19—22 days (and young fed only 3 weeks) by both parents. Diet: chiefly crustaceans and small fish. Migrant. Migration: IV—VI and VIII—IX.

3 **SANDWICH TERN** *(Sterna sandvicensis)* Size of Black-headed Gull, with long, slim, black, yellow-tipped beak. Long, erectile black feathers on back of head. Black, relatively long feet. Deeply forked tail. Winter plumage: white forehead, grey crown. Juvenile: black and white crown, dark-spotted back, yellowish beak. Voice: frequent, energetic 'kirrik', reminiscent of Partridge. Widespread nester on sandy coasts and flat, sandy islands, occurs inland as accidental bird only. Breeding season: V—VI, 1 brood a year. Nest: shallow, straw-lined depression in sand. Eggs: 2—3, sandy (sometimes very pale) with brownish black spots, incubated 22 days by both parents. Diet: small sea-fish; often hovers, then dives after prey. Migrant. Migration: IV—V and VII—IX.

1

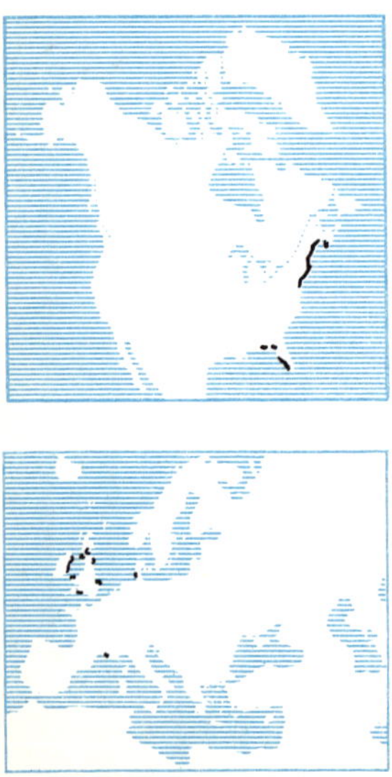

2

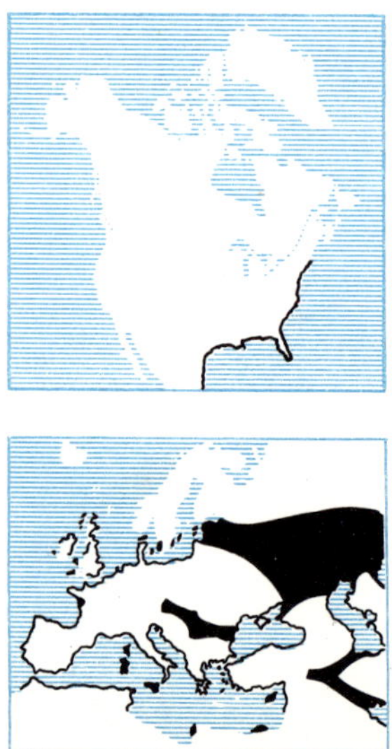

3

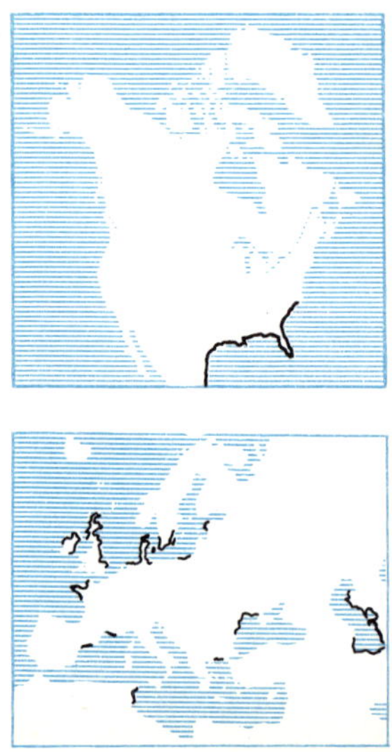

1 S
1 S
1 imm
2 S
2
2 D
2 imm
3 S
3 S
3 W
3 imm

1 **RAZORBILL** *(Alca torda)* Smaller than Mallard. Black and white; large head, short neck (unlike divers and ducks). In flight easily recognized by stocky body, short, triangular tail and short, almost whirring wings. Sits high in water, dives frequently. Distinguished from other auks by shape of black, white-lined, flat-sided beak. ♂ = ♀. Winter and juvenile plumage: throat and front of neck white. Downy plumage: back brown, head and underside white. Voice: deep, croaking sounds and soft, rippling whistle. Nests in colonies on rocky north-west European coasts. Breeding season: IV—VI. Egg: 1, pear-shaped, grey with black spots, laid in niche on bare rock, incubated 32—36 days (and young fed 1 month) by both parents. Diet: fish, crustaceans. Partial migrant. Migration: II—III and IX.

2 **PUFFIN - COMMON PUFFIN** *(Fratercula arctica)* Slightly smaller than pigeon. High-ridged, triangular, variegated beak, large pale grey-sided head, vermilion legs. Winter plumage: beak yellowish brown, sides of head dark grey. Juvenile: black, less prominent beak. Downy plumage greyish black, belly white. Flies and swims like Razorbill, stands erect. Voice: growling 'arr'. Nests in colonies on rocky coasts and steep, rocky islands in sea. Breeding season: V—VI. Nests in long burrow dug in turf. Egg: 1, pale grey, unspotted, incubated 33—37 days (and young fed at least 40 days) by both parents. Diet: small sea-fish, crustaceans, molluscs, worms. Partial migrant. Migration: III—IV and VIII—IX.

3 **GUILLEMOT - COMMON MURRE** *(Uria aalge)* About size of Mallard. Back black, underside white. Pointed beak like diver's. Often has white stripe behind white-ringed eyes. Beak and legs black. Winter and juvenile plumage: cheeks, throat and front of neck white. Downy plumage: back blackish grey, underside white. Swims and dives like Razorbill, sits on whole of legs. Voice: loud cries, 'arrr' or 'arra'. Widespread colony nester on cliffs on rocky coasts. Breeding season: V—VI. Egg: 1, pear-shaped, variably coloured (stone grey to bluish green, with dark spots and scribbles), incubated 28—31 days by both parents. Parents protect and feed young about 20 days on cliff, afterwards also on sea. Diet: mainly small sea-fish, fewer crustaceans, molluscs, worms. Partial migrant. Migration: III—IV and VIII—IX.

1

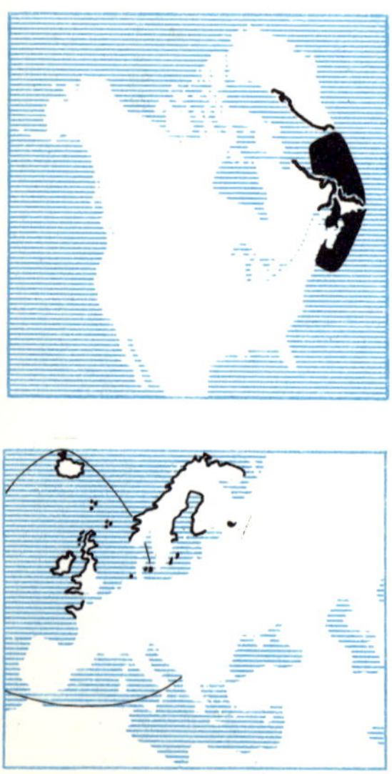

2

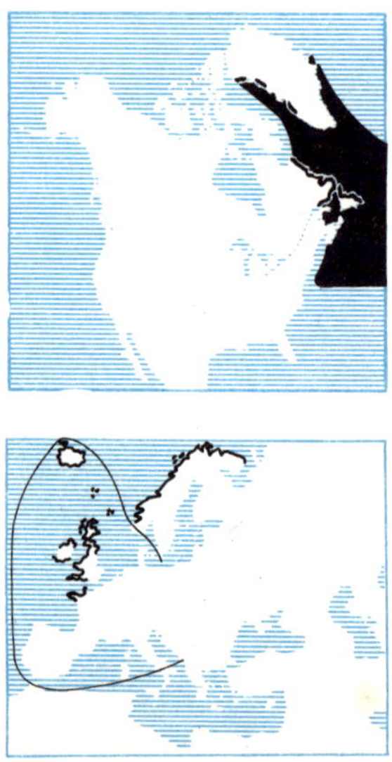

3

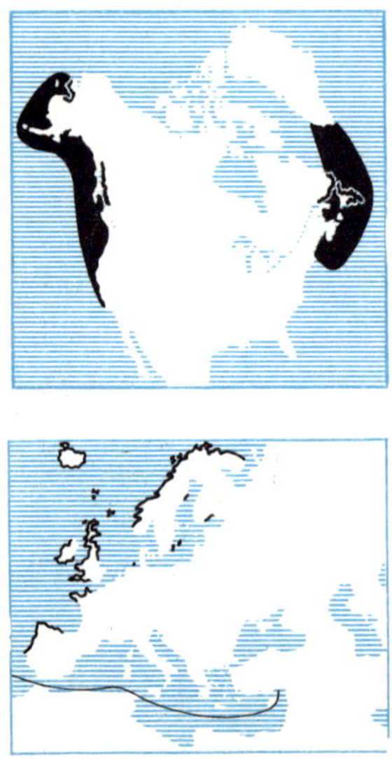

1 S
1 S
1 juv
2 juv
2 S
3 juv
3 S
3 S

Order: Auks - Alciformes **Family: Auks** - Alcidae

1 **BLACK GUILLEMOT** *(Cepphus grylle)* About size of pigeon. Black, with distinctive white patch on wings. Beak black, legs red. Winter plumage more or less white on underside. Juvenile similar to winter plumage, but with white patch on wings and black-spotted underside. Swims with head raised, dives and sits like Razorbill. Downy plumage blackish brown. Voice: feeble, whistling 'peeee'. Nests on rocky coasts and islands, mostly singly or in small colonies. Breeding season: V—VI. Eggs: 2, grey or brown, dark-spotted, laid directly in cavity in ground, incubated 27—30 days (and young fed at least 35 days) by both parents. Diet: mainly small sea-fish, some crustaceans and worms. Migrant. Migration: III—IV and IX—X.

Order: Pigeons and Doves - Columbiformes **Family: Pigeons and Doves** - Columbidae

2 **ROCK DOVE** *(Columba livia)* Size of domestic pigeon. Bluish grey, with white rump, 2 black wing bars. Juvenile browner. Downy plumage thin, creamy yellow. Domestic pigeon populations living wild in towns are birds which have reverted over generations to form resembling Rock Dove. Voice: cooing 'ōōr-ōōr-rōō-cooo'. Nests on cliffs, mostly on Mediterranean coast, rarely inland. Breeding season: IV—VII, 2—3 broods a year. Nest: pile of plant fragments in rock crevice or niche. Eggs: 2, white, incubated 17—18 days, mainly by ♀. Young reared 4—5 weeks by both parents. Diet: chiefly seeds, grain, also snails and slugs. Resident bird.

3 **STOCK DOVE** *(Columba oenas)* About size of domestic pigeon. Forest-dweller. Without white rump and white wing and neck spots. Back light bluish grey, 1—2 indistinct black wing bars. Tail grey, with indistinctly defined black band at end. Sides of neck iridescent green; claret, brown-tinged crop. Juvenile: brown-grey back, no lustre on sides of neck. Downy plumage like Wood Pigeon's. Voice: wailing, rather than cooing 'hōōh-hōō-hoo-hoo'. Nests in old, open woods. Breeding season: IV—VII, 2—3 broods a year. Nest: pile of twigs, stalks and leaves in tree hollow. Eggs: 2, white, incubated 17—18 days by both parents. Care of young about 25 days. Diet: seeds. Partial migrant. Migration: III and VIII—X.

1

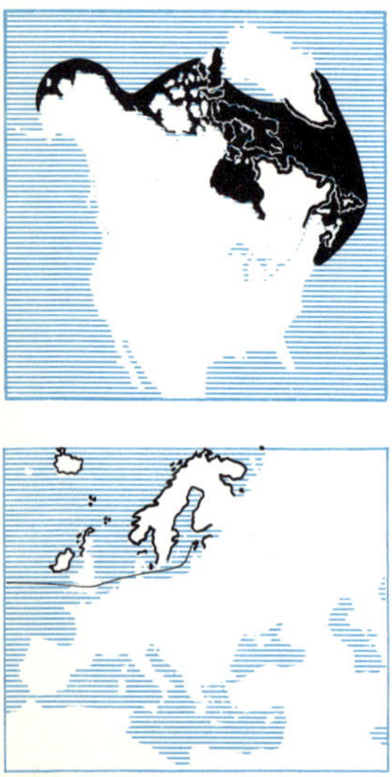

2

3

1 S
1 juv
1 S
2
3

Order: Pigeons and Doves - Columbiformes **Family: Pigeons and Doves -** Columbidae

1 **WOOD PIGEON** *(Columba palumbus)* Size of large domestic pigeon. Back dark greyish brown, rump bluish grey, white spot on side of neck. White anterior wing border forms distinctive white cross-band on upper surface of wings in flight. Juvenile has no neck spots. Flies fast, often claps wings when rising and during undulating display flight. Voice: cooing 'coo-coo-roo, coo-coo'. Common in all types of woods, in western Europe also occurs in towns. Large flocks frequently seen in fields and meadows towards end of summer. Breeding season: IV—VIII, 2—3 broods a year. Nest: shallow dish, made of twigs, in tree. Eggs: 2, white, incubated 16—17 days (and young fed some 4 weeks) by both parents. Diet: seeds, beechnuts, acorns, some berries and snails. Partial migrant. Migration: II—III and IX—XI.

2 **TURTLE DOVE** *(Streptopelia turtur)* Smaller and daintier than domestic pigeon. Long, rounded tail often fanned in flight, showing white tips of tail feathers. Back rusty brown, crop and breast claret. Juvenile: browner head, back and wings. Downy plumage fawn, thin. Flies with jerky wing strokes. Voice: serial throbbing 'rrōōrrrr rrōōrrrr rrōōrrrr'. Widespread nester in open country with bushes, hedges, groups of trees and copses. Breeding season: V—VII, 1 brood a year. Shallow nest made of a little brushwood in bush. Eggs: 2, white, incubated 14—16 days by both parents. Young fed about 3 weeks. Diet: seeds. Migrant. Migration: IV—V and VIII—IX.

3 **COLLARED TURTLE DOVE** *(Streptopelia decaocto)* Larger, more robust than Turtle Dove, with longer tail. Fawny brown, with black nuchal ring. Dark primaries, broad white band at end of tail. ♂ = ♀. Juvenile: browner on underside, nuchal band very narrow or absent. Downy plumage like Turtle Dove's. Voice: throaty 'coo-cooo, coo', with second syllable accented, pursuit call 'hwee'. Widespread, common bird in gardens and parks. Breeding season: III—X, 2—4 broods a year. In display flight, rises obliquely, glides down with spread wings. Nest: flat pile of brushwood on tree or building. Eggs: 2, white, incubated 14—16 days. Young leave nest at 14—20 days, still fed 3 more weeks. Diet: seeds, berries, parts of plants. Resident bird.

1

2

3

1 ad
1 ad
1 juv
2
2 ad
3
2 juv
3 D

Order: Owls - Strigiformes **Family: Barn Owls** - Tytonidae

1 **BARN OWL** *(Tyto alba)* About size of pigeon; large head, heart-shaped face. Whitish grey, with white beading; underside white to rusty yellow, with dark brown spots. Dark eyes, long, slim wings. Rather long legs covered with very short feathers. ♂ = ♀. Juvenile: same as summer plumage. Downy plumage white. Voice: loud, trailing 'khree-i' when flying, peculiar snoring sounds on nest. Widespread nester in human habitations (in church towers, castles, barns, etc.). Hunts in open country. Breeding season: IV—VIII, 1—2 broods a year. Nest: in dark corner of loft, without bedding. Eggs: 4—7 or more, white, incubated 30—34 days (and young fed 7—9 weeks) by both parents. Diet: small mammals, fewer birds and insects. Resident bird.

Order: Owls - Strigiformes **Family: Owls** - Strigidae

2 **SNOWY OWL** *(Nyctea scandiaca)* Size of Eagle Owl. Large, round head, amber eyes. Fur-like plumage mainly white. ♂ pure white, or with a few dark cross-bands, ♀ more richly banded. Juvenile: largely greyish brown, with black spots and cross-bands. Downy plumage white. Wide, rather short wings, long, triangular tail. Diurnal bird. Voice: seldom heard, sometimes croaking sounds or serial 'rick rick rick'. Nests in arctic tundras, almost regular winter bird in northern Europe, accidental bird further south. Breeding season: IV—VI, 1 brood a year. Nest: unlined depression in ground. Eggs: 7—9, rounded, white, incubated 32—34 days by ♀. Young fed 50—60 days by both parents. Diet: mainly small mammals (lemmings, voles), fewer birds. Partial migrant, occurs in western and central Europe from X to I.

3 **EAGLE OWL** *(Bubo bubo)* Largest owl, much larger than buzzard. Wing span 170 cm. Large head with movable ear-tufts, orange red eyes. ♂ = ♀ (slight difference in size). Juvenile less vividly marked than adults. Downy plumage yellowish white. Flies silently, alternately 'rowing' and gliding. Nocturnal. Voice: mating call deep, repeated 'bōō-hōō'. Rare nester in large forests at low and high altitudes, mountain ravines and rocky river valleys. Breeding season: III—V. Nest: usually in hollow in rocky slope or rock niche, rarely in tree (old eagle or heron nest). Eggs: 2—3, white, almost spherical, incubated about 15 days by ♀ (♂ fetches food). Young fed and tended 2½ months by both parents. Diet: small and moderately large mammals and birds. Resident bird.

1

2

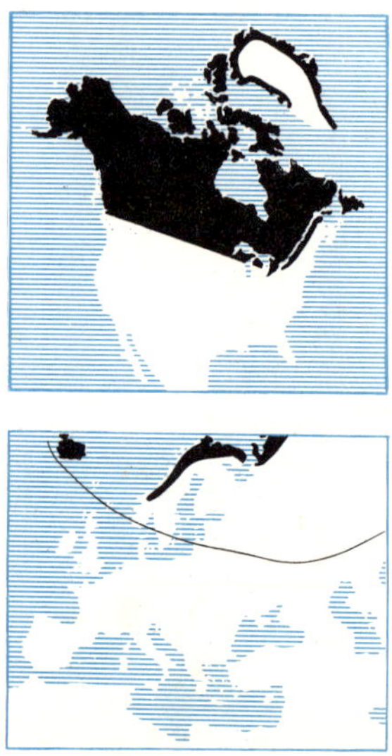

3

1 LV
1
1 DV
1 D
2 ♂ LV
2 ♀ DV
3
3 D

1 **LONG-EARED OWL** *(Asio otus)* About size of pigeon. Round face, white facial disc feathers form X between eyes. Long ear-tufts, yellow-red eyes. ♂ = ♀ = juvenile. Downy plumage white, with black-ringed eyes and visible ear-stubs (difference from Tawny Owl). Flies noiselessly; distinguished by large head, long, slender wings and slow wing strokes. By day rests in tree, often beside trunk, hunts after dusk. Voice: muffled 'ōō', uttered in slow rhythm; young birds, after leaving nest, persistently screech 'psheea'. Widespread forest-dweller, hunts along margin of woods. In winter often roosts in large parties. Breeding season: IV—VI, 2 broods a year. Nest: in tree, usually old nest of bird of prey or crow. Eggs: 4—7, white, rounded, incubated 27—28 days by ♀. Diet: chiefly voles. Partial migrant. Migration: III and X—XI.

2 **SHORT-EARED OWL** *(Asio flammeus)* About size of Long-eared Owl. Yellowish brown, appears pale. Ear-tufts short, usually folded and invisible. Round, white facial disc, yellow eyes set in black mask. Long, narrow wings with very pale under surface. Rocking flight with slow strokes. Often active by day. ♂ = ♀ = juvenile. Downy plumage cream, later with round, black mask and white 'moustache'. Voice: rapidly repeated 'boobooboo' calls during display flight, also claps wings hurriedly when swooping. Irregular, infrequent nester in wet, open country, e.g. moors, wet meadows. When migrating, appears in parties in potato fields, willow beds, etc. Breeding season: IV—V, 1 brood a year. Nest: on ground, concealed in grass or reeds, made of miscellaneous plant parts. Eggs: 4—7, white, incubated 26—27 days by ♀. Diet: mainly fieldmice. Partial migrant. Migration: III—IV and IX—XI.

3 **LITTLE OWL** *(Athene noctua)* About size of Blackbird. Large, flat-topped head, wide oval face, yellow eyes. Short tail and short-feathered, rather long legs (cf. Tengmalm's Owl). Small white droplet spots on back, longitudinally striped breast. ♂ = ♀. Juvenile browner. Downy plumage white. Often sits huddled up, bobs up and down when excited. Voice: display call protracted 'ōōk', frequently calls 'ividd' and excited, yapping 'kiew'. Often active by day. Short, rounded wings, fluttering, undulating flight. Fairly common in open country with trees, often nests in human communities. Breeding season: IV—V, 1 brood a year. Nest: hole in wall or tree, in loft, etc., without bedding. Eggs: 4—6, white, rounded, incubated 28 days by ♀. Diet: murine rodents, insects, occasionally birds. Resident bird.

1

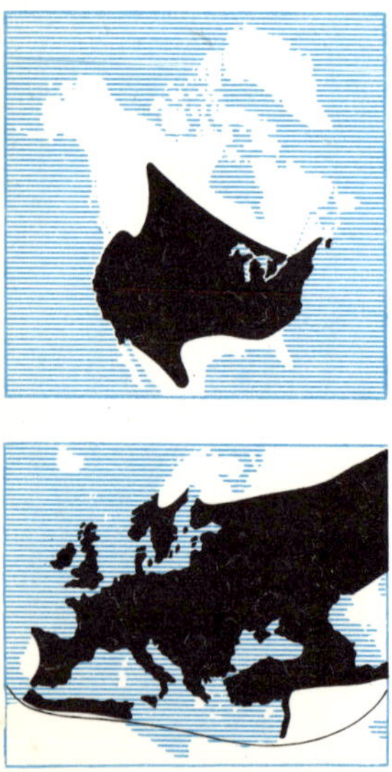

2

3

1
1 D
2
2
2 D
3
3 D
3

1 **PYGMY OWL** *(Glaucidium passerinum)* Smallest owl, no larger than Starling. Inhabits deep forests. Wide oval face, light yellow eyes, white 'eyebrows'. Twitches short tail when excited. ♂ = ♀ = juvenile. Downy plumage white. Undulating flight, often seen by day. Not timid. Voice: monosyllabic, repeated 'whee' reminiscent of Bullfinch's whistle, also serial 'deew, ew-ew-ew'. Rare nester in old mountain pine-woods. Breeding season: IV—V, 1 brood a year. Nest: in tree hollow (woodpecker's hole). Eggs: 4—6, white, incubated 28 days by ♀. Care of young about 6 weeks. ♂ fetches food (small mammals and songbirds). Resident bird.

2 **SCOPS OWL** *(Otus scops)* Almost size of Blackbird. Grey-brown, with bark-like markings, short ear-tufts like horns, lemon yellow eyes. ♂ = ♀ = juvenile. Nocturnal. Flies quite fast and skilfully, has relatively long wings and short tail. Voice: display note persistent, monotonous 'kyew'. South European bird, nests in open country with old trees, often in orchards near human habitations. Breeding season: V—VII, 1 brood a year. Nest: in hole in tree or wall. Eggs: 3—5, white, incubated 24—25 days by ♀. Young fed about 5 weeks by both parents. Diet: mainly large insects, fewer mice and birds. Migrant. Migration: IV and VIII—X.

3 **TAWNY OWL** *(Strix aluco)* Larger and more robust than Long-eared Owl. Large, round face, round facial disc, brownish black eyes, short legs feathered down to toes. Grey and rust brown colour variant. ♂ = ♀ = juvenile. Downy plumage white. Nocturnal. Noiseless flight; wide, rounded wings, short tail. Voice: wailing display call with two phrases, 'hōō, oo-hooooooo', and shrill 'ke-wick'. Common forest bird, also found in large parks. Breeding season: III—IV, 1—2 broods a year. Nest: mostly in hollow tree, occasionally in building or hole in ground. Eggs: 3—4, white, rounded, incubated 28—30 days by ♀. Young remain 4—5 weeks in nest, still supplied with food by both parents at least 3 weeks after leaving nest. Diet: mainly mice, other small mammals, songbirds, large insects. Resident bird.

1

2

3

1 juv
1
2 grey form
2 brown form
2 D
3 brown form
3 grey form
3 D

1 **URAL OWL** *(Strix uralensis)* Larger than Tawny Owl, with longer tail. Mostly very pale, but dark (brown) variant also exists. Round face, black-brown eyes. ♂ = ♀ = juvenile. Downy plumage white. Distinguished in flight by wide, fairly long wings and longish tail. Active before nightfall. Voice: wild 'hoo, wow-wow-wow' and hoarse 'kawveek'. Rare nester in old (especially beech and mixed) woods. Breeding season: III—V, 1 brood a year. Nest: in tree hollow, without bedding, sometimes nest of bird of prey. Eggs: 3—4, white, incubated 27—29 days by ♀. Young fed by both parents about 5 weeks in nest and several weeks after leaving nest. Diet: mice, birds, insects. Resident bird.

2 **TENGMALM'S OWL - BOREAL OWL** *(Aegolius funereus)* Slightly larger than Little Owl, with rounded head and round facial disc. Usually sits erect, has long tail and long-feathered legs (cf. similar Little Owl). Large, light droplet spots on back. ♂ = ♀. Juvenile almost entirely coffee-coloured. Downy plumage brownish white. Flies in straight line (cf. Little Owl), active only at night. Voice: tuneful, vibrating 'poo-poo-poo-poo-poo' reminiscent of steam-engine whistle. Nests in old woods at all altitudes. Breeding season: III—V, 1 brood a year. Nest: in tree hole (usually woodpecker's hole), no bedding. Eggs: 4—6, white, incubated 26—27 days by ♀. Young tended 7—8 weeks by both parents. Diet: small mammals, songbirds. Resident bird.

3 **HAWK OWL** *(Surnia ulula)* Larger than Long-eared Owl. Black-framed, square face, flat-topped head, light yellow eyes, long, cross-banded tail and brown and white ripple-marked underside. ♂ = ♀. Juvenile: more grey-brown, with no bands on under tail-coverts. Downy plumage white. Resembles bird of prey (especially Sparrow Hawk) in flight (long tail, short wings, underside). Starts hunting before nightfall. Voice: mating call monotonous serial 'oo-oo-oo-oo-oo'. Nests in Scandinavian taiga forests, in winter sometimes irruption bird in northern and (less often) central Europe. Breeding season: III—IV. Nest: in tree hole (woodpecker's hole) or old nest of bird of prey. Eggs: 4—7, white. Little known of nest biology. Diet: mice, lemmings, songbirds. Winter bird from X to II.

1

2

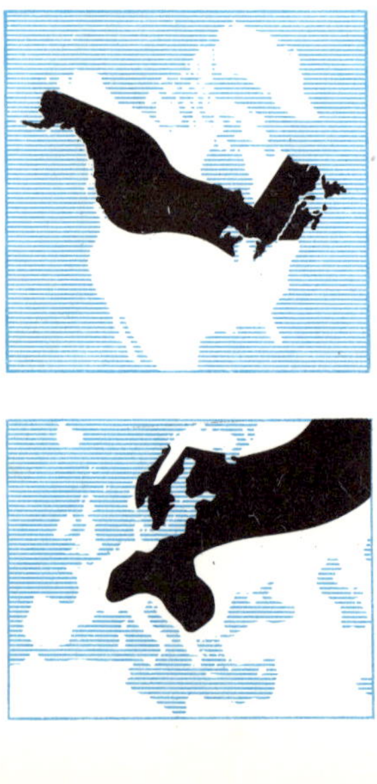

3

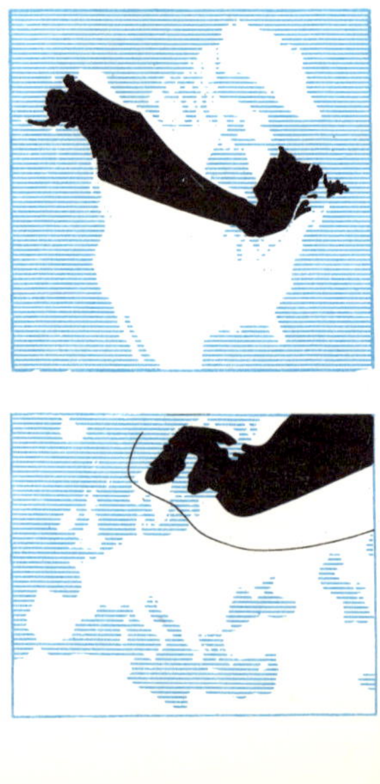

1 D
1
2 juv
2
2
3
3

Order: Cuckoos - Cuculiformes | **Family: Cuckoos** - Cuculidae

1 **CUCKOO** *(Cuculus canorus)* Size of Turtle Dove. ♂ grey, with cross-banded belly, ♀ brown on underside (sometimes completely reddish brown), with cross-banded belly. Juvenile: rusty white scales on back. Nestling at first naked, with dark violet skin and orange red gullet. Like Sparrow Hawk or Kestrel in flight, but flatter wing strokes. Voice: call of ♂ tuneful 'cuc-coo', often repeated three times, hoarse pursuit call 'wow-wow-wow'; ♀ a rapid bubbling note, 'quickquickquickquick'. Inhabits all types of woods, park and meadows with groups of trees. Nest parasite, lays 15—20 highly variable eggs, singly, in nests of insectivorous songbirds. Breeding season: V—VII, incubation period 12½ days, hatching period 21—23 days, young fed 3 more weeks. Diet: insects. Migrant. Migration: IV—V and VII—VIII.

Order: Nightjars - Caprimulgiformes | **Family: Nightjars** - Caprimulgidae

2 **NIGHTJAR** *(Caprimulgus europaeus)* About size of Blackbird. Large head, short beak, large eyes, short legs, bark-like markings. ♂ has white wing spots and corners of tail (not ♀). Juvenile like ♀. Downy plumage bluish grey. Distinguished in flight by long, narrow wings and long tail. Crouches along branch by day. Voice: in evening and at night a loud, rapid, churring song, 'urrurrurrurr', etc., claps wings loudly when flying. Nests in dry conifer woods, on heaths, forest paths, margin of woods. Breeding season: V—VII, 2 broods a year. Eggs: 2, elongate, brown-mottled, laid on bare ground, incubated 17—20 days by both parents. Diet: nocturnal insects, especially moths. Migrant. Migration: IV—V and VIII—IX.

Order: Rollers and Allies - Coraciiformes | **Family: Hoopoes** - Upupidae

3 **HOOPOE** *(Upupa epops)* Larger than Blackbird. Orange brown, black and white-chequered plumage, long, thin, curved beak and erectile, fan-like crest are unmistakable. ♂ = ♀ = juvenile. Downy plumage greyish white. Flutters like butterfly, effect very colourful. Voice: display call trisyllabic 'hoop-poo-poo', pursuit call hoarse 'chāārr'. Inhabits pasture and meadow country with willows. Breeding season: V—VII, 1—2 broods a year. Nest: in hole, poorly lined (mostly in tree, wall, pile of stones or brushwood, sometimes under roof). Eggs: 6—7, grey, finely speckled, incubated 16 days by ♀. Hatching period about 4 weeks, young fed 1 more week by both parents after leaving nest. Diet: insects, especially mole crickets. Migrant. Migration: IV and IX—X.

1

2

3

1 ♂
1
1 ♀ brown phase
2
2 ♂
3
3

1 **ALPINE SWIFT** *(Apus melba)* Considerably larger than Swift. Greyish brown, with brown breast band, throat and belly white. ♂ = ♀. Juvenile has light-edged feathers. Nestling without plumage at first. Rapid, skilful flight; long, narrow, curved wings, short, forked tail. Voice: loud, trilling 'tirrr-tree-tree-tree' while flying. Inhabits southern Europe, nests in rocky mountains or under roof of tall buildings in towns. Breeding season: V—VI, 1 brood a year. Nest: under roof or in crevice, shallow, grass blade and straw fragments stuck together, with a few feathers. Nests in colonies. Eggs: 2—3, white, incubated 20 days, mainly by ♀. Young fed 45—50 days by both parents. Diet: flying insects. Migrant. Migration: III—IV and VIII—IX.

2 **SWIFT** *(Apus apus)* Black, swallow-like, with narrow, curved wings and fast, whirring flight. Circles, mostly in numbers, round church towers and other tall buildings. Able to cling to vertical walls. ♂ = ♀. Juvenile has light-edged feathers on head and back. No downy plumage. Voice: screaming 'sweeer sweeree', mostly in chorus. Widespread nester in human habitations. Breeding season: V—VI, 1 brood a year. Nest: straw, hairs and feathers stuck together, in dark corner of loft in tall building. Nests in colonies. Eggs: 2, white, incubated 18—20 days by both parents. Young fed about 50 days. Diet: flying insects. Migrant. Migration: V and VIII.

Order: Rollers and Allies - Coraciiformes **Family: Rollers** - Coraciidae

3 **ROLLER** *(Coracias garrulus)* About size of Jackdaw. Gaily coloured, azure with red-brown back. Large beak, moderately long, straight-ended tail, shortish legs. Flies like Jackdaw, courting ♂ flutters. Sits on pylons, corn-sheaves, etc., watching for insects. ♂ = ♀. Juvenile: greenish brown instead of blue. No downy plumage. Voice: raucous 'krak-ak-ak', etc. Rare and sporadic summer bird, common only in eastern and southern Europe (there often in colonies). Breeding season: V—VII, 1 brood a year. Nest: in hollow tree (mostly Black Woodpecker's hole), lined with a little vegetation and a few feathers. Eggs: 4—5, white, incubated 19 days. Young fed in nest by both parents up to 50 days. Diet: mainly large insects, fewer mice and lizards. Migrant. Migration: IV—V and VIII—IX.

1

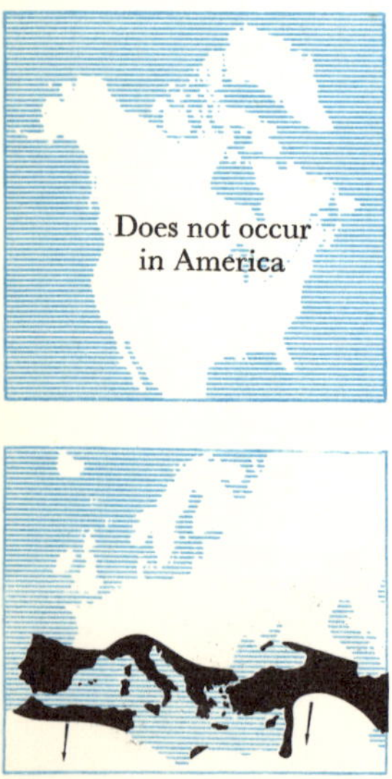

2

3

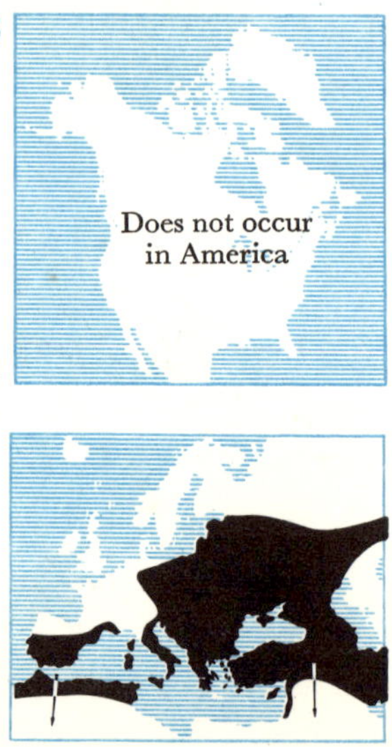

2 juv
2
2
1
3 juv
3

Order: Rollers and Allies - Coraciiformes **Family: Kingfishers** - Alcedinidae

1 **KINGFISHER** *(Alcedo atthis)* Size of large Sparrow. Large head, dagger-like beak, short tail. Iridescent blue-green back, russet underside. Form and colouring unmistakable. ♂ = ♀ = juvenile. No downy plumage. Darts over water, with whirring wings, uttering loud, shrill 'chee-chee'. Widespread, but not common, found by clean rivers, streams and lakes. Breeding season: IV—VII, 2 broods a year. Nest: at end of horizontal burrow dug in steep, clayey or sandy bank. Eggs: 6—7, spherical, incubated 21 days by both parents. Young fed about 25 days in nest and only few days after leaving nest. Diet: mainly small fish, also aquatic insects, caught in vertical dive. Partial migrant, no fixed migration time.

Order: Rollers and Allies - Coraciiformes **Family: Bee-eaters** - Meropidae

2 **BEE-EATER** *(Merops apiaster)* About size of thrush, gaily coloured. Underside blue, throat yellow, back chestnut and yellow. Thin, sharp beak same length as head; long, pointed tail feathers. Flight like Swallow's, often interspersed with gliding. ♂ = ♀. Juvenile's tail minus point. Downy plumage not developed. Voice: constantly repeated 'prruik' while flying. Nests in southern Europe, often in large colonies. Breeding season: VI—VII, 1 brood a year. Nest in burrow in steep bank, sandpit, etc. Eggs: 5—6, white, spherical, incubated about 20 days by ♂ and ♀. Eggs: 5—6, white, spherical, incubated about 20 days by ♂ and ♀. Young fed 20—25 days in nest and a few days outside. Diet: flying insects, especially hymenoptera, butterflies and dragonflies. Migrant. Migration: V and VIII—IX.

Order: Woodpeckers - Piciformes **Family: Woodpeckers** - Picidae

3 **WRYNECK** *(Jynx torquilla)* Size of Sparrow. Bark-coloured, non-constrasted markings, small beak, non-supporting tail. Feet adapted for climbing (2 toes point forwards, 2 backwards), but does not climb vertical trees. ♂ = ♀ = juvenile. No downy plumage. Voice: series of 8—12 prolonged, rising 'quee-quee-quee' notes. Widespread, inhabits open country with groups of trees and orchards. Breeding season: V—VI, 1 brood a year. Nest: in tree hole, not self-made (cf. other woodpeckers). Eggs: 7—10, white, incubated 13—14 days by both parents. Young remain in nest about 25 days. Diet: mainly ants and ant larvae, less often other insects. Migrant. Migration: IV—V and VIII—IX.

1

2

3

1
1
2
3

Order: Woodpeckers - Piciformes **Family: Woodpeckers -** Picidae

1 **BLACK WOODPECKER** *(Dryocopus martius)* Size of crow. All black, with dark red crown (♂) or nape (♀). Juvenile more brownish black. Downy plumage not developed. Flies in straight line, not undulatingly like other woodpeckers. Voice: loud, carrying 'kleea', followed by 'krri-krri-krri-krri'; frequent sonorous 'choc-choc-choc' when flying. Drums powerfully, with a long, slow roll. Widespread nester in large forests with old timber. Breeding season: IV—V, 1 brood a year. Nest: deep hole with oval entrance, excavated in tree. Eggs: 4—5, white, incubated 12—14 days by both parents. Young fed about 25 days in nest and a few days outside. Diet: beetle larvae living in wood, ants. Resident bird.

2 **GREEN WOODPECKER** *(Picus viridis)* Size of pigeon. Mainly olive green, with red crown. Shows yellow rump and barred primaries in flight. ♂ has red, ♀ black, moustachial stripe. Juvenile distinguished by dull red cap and spotted underside. No downy plumage. Undulating flight. Spends most of time on tree trunks, but sometimes looks for food on ground. Voice: mating call a loud, laughing 'queu-queu-queu-queu'. Common nester in open woods, copses, parks, orchards. Breeding season: IV—V, 1 brood a year. Nest: hole hacked in tree, with roughly round entrance. Eggs: 5—7, white, incubated 15—17 days by both parents. Parents feed young about 20 days in nest and lead them for a further 14 days. Diet: mainly ants and ant larvae, fewer other insects. Resident bird.

3 **GREY-HEADED WOODPECKER** *(Picus canus)* Smaller than Green Woodpecker. Greyish green, with slate grey neck and head. Only ♂ has bright red cap. Juvenile: spotted belly and tail. Downy plumage not developed. Resembles Green Woodpecker in flight. Voice: laughing call, slower and more trailing than Green Woodpecker's, often composed of only 3—4 notes. Sporadic nester in hills and mountains, prefers park-land, orchards and open country. Breeding season: V—VI, 1 brood a year. Nest, eggs and care of brood largely as for Green Woodpecker. Diet: chiefly ants, also other insects. Resident bird.

1

2

3

1 ♂
1 ♂
1 ♀
2 ♂
2 ♀
2 juv
3 ♀
3 ♂
2 ♂

1 **GREAT SPOTTED WOODPECKER** *(Dendrocopos major)* Size of Starling, good climber. Black and white, chequered plumage. Distinguished by large, black moustachial stripe stretching to neck and deep red under tail-coverts standing out sharply against white belly. ♂ has red nape, ♀ black crown and nape, juvenile red crown. Downy plumage absent. Flies in long curves. Voice: mostly a loud, hard 'tchick', strung together to 'kik-kik-kik-kik' when excited. Regular drumming. Nests in all types of woods and in parks. Breeding season: V—VI, 1 brood a year. Nest: hole excavated in tree, round entrance. Eggs: 5—7, white, incubated 12—13 days by both parents. Parents feed nestlings some 3 weeks and a short time after young leave nest. Diet: insects living in wood, conifer seeds, fruit, berries. Partial migrant and resident bird.

2 **SYRIAN WOODPECKER** *(Dendrocopos syriacus)* Closely resembles Greater Spotted Woodpecker, but lacks band under cheek. Under tail-coverts brighter red. ♂ has red nuchal spot, ♀ not. Juvenile: red cap, often reddish breast. No downy plumage. Voice: softer than Greater Spotted Woodpecker's, calls 'tchick' and 'tchirruck'. Inhabits southern Europe, advancing into central Europe, very occasionally nests there. Frequents open country, especially orchards and avenues. Breeding season: V—VI, 1 brood a year. Nest: hole excavated in tree, round entrance. Eggs: 4—5, incubated 14—15 days by both parents. Parents feed nestlings 17—21 days, with berries, fruit and nuts far more than with insects. Resident bird.

3 **MIDDLE SPOTTED WOODPECKER** *(Dendrocopos medius)* Slightly smaller than Greater Spotted Woodpecker. ♂, ♀ and juvenile all have red crown not separated from cheeks by black line (cf. 1 and 2). Pink under tail-coverts not sharply defined from whitish belly. Isolated black streaks on underside especially marked in juvenile. Thin beak. No downy plumage. Voice: call-note soft 'kik' or 'ptik', in spring hoarse, croaking scream of 6—9 'wait' syllables. Seldom drums. Isolated nester in deciduous woods. Breeding season: IV—VI, 1 brood a year. Nest: hole excavated in tree, round entrance. Eggs: 5—6, white, incubated 12 days. Young fed at least 14 days in nest and about 1 week after leaving nest. Both parents share in care of brood. Diet: beetles living in wood, ants, pine seeds. Resident bird.

1

2

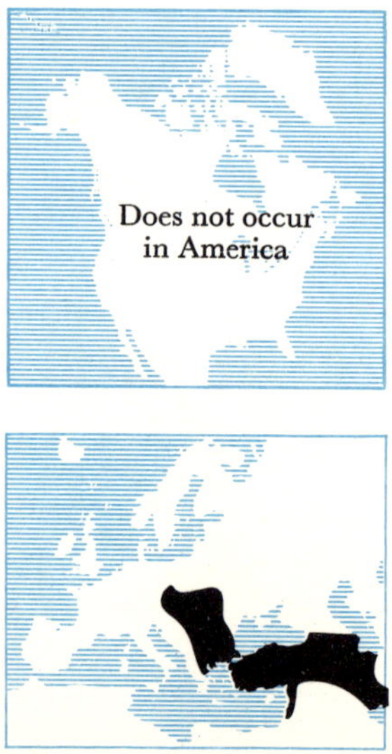

3

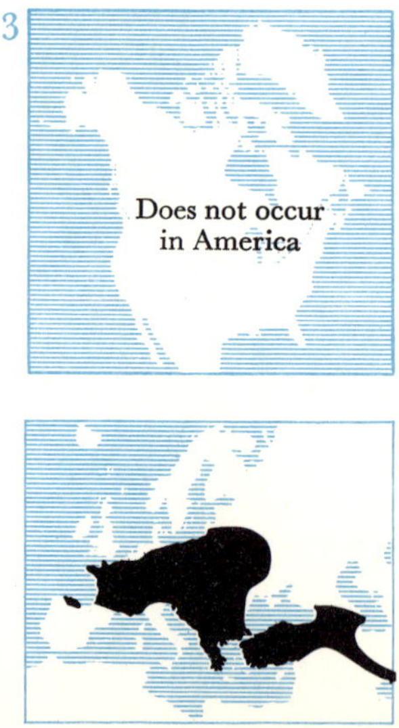

1 ♂
1 juv
1 ♀
1 ♂
2 ♂
3 ♀
3 ♂

1 **WHITE-BACKED WOODPECKER** *(Dendrocopos leucotos)* Larger and slimmer than Greater Spotted Woodpecker, with longer, thinner beak. No white scapulars, distinguished by white-ended back and white rump (cf. birds on p. 152). ♂ has red, ♀ black, crown. Under tail-coverts red, not sharply defined from white underside. Juvenile: little red on head and belly. Downy plumage not developed. Voice: call-note 'tchick', softer than Greater Spotted Woodpecker's; ♂ and ♀ both drum. Nests in old timber in mixed and deciduous woods, especially in mountains. Breeding season: IV—V, 1 brood a year. Nest: large hole excavated in tree. Eggs: 4—5, white, hatching period 27—28 days. Both parents tend brood. Diet: beetles living in wood, ants. Resident bird.

2 **LESSER SPOTTED WOODPECKER** *(Dendrocopos minor)* Size of Sparrow. Vivid black and white. No white shoulder spots, but entire back thickly marked with white cross-bands. ♂ has red crown, ♀ not. Underside completely white. Juvenile: brownish underside, striped sides. Downy plumage not developed. Voice: clear, serial 'kee-kee-kee-kee'; drums often. Widespread in flat, open country with groups of trees, avenues, orchards. Breeding season: IV—V, 1 brood a year. Nest: hole excavated in tree, round entrance. Eggs: 5—6, white, incubated 11 days. Young fed about 20 days in nest and 10 more days in open. Both parents care for brood. Diet: mostly beetles and beetle larvae, fewer ants, in winter also seeds. Resident bird.

3 **THREE-TOED WOODPECKER — NORTHERN THREE-TOED WOODPECKER** *(Picoïdes tridactylus)* Somewhat smaller than Greater Spotted Woodpecker. Broad black cheek stripes, no red. Back entirely white, with a few black spots; white scapulars absent, vivid black bands on sides. ♂ has lemon yellow, ♀ grey, crown. Juvenile: generally greyer, white back more spotted. No downy plumage. Voice: call-note a chattering 'kek-ek-ek-ek', drums slowly. Inhabits old forests in mountains, in north also in plains. Breeding season: V—VI, 1 brood a year. Nest: hole excavated in tree, round entrance. Eggs: 3—4, white. Both parents incubate eggs and feed young, length of incubation and hatching period not yet known. Diet: mainly beetles, butterflies, ants. Resident bird.

1

2

3

1 ♀
1 ♂
1 ♂
2 ♂
2 ♀
2 ♂
3 ♂
3 ♀
3 ♂

1 **SKYLARK** *(Alauda arvensis)* Larger, more robust than Sparrow. Umber, black-striped. Outer tail feathers white. Crown feathers ruffled from time to time. Runs on ground, crouches if in danger. ♂ = ♀. Juvenile: somewhat darker, with white-edged feathers on back. Downy plumage: sparse, long, straw-coloured. Voice: display song sung while fluttering high above ground, often for several minutes; call-note a quavering 'chir-r-up'. Common in cultivated grasslands (fields, meadows), occurs sporadically on broad mountain ridges. Breeding season: IV—VII, 2 broods a year. Nest: in depression in ground, loosely woven structure made of grass blades and roots. Eggs: 3—5, greyish brown, thickly brown-speckled, incubated 12—14 days by ♀. Young leave nest before fully fledged, fed by both parents. Diet: seeds, green parts of plants, insects. Partial migrant. Migration: II and IX—X.

2 **WOOD LARK** *(Lullula arborea)* Smaller than Skylark. Short tail, outer tail feathers brown, light-tipped. Light eye stripes extending to nape. ♂ = ♀. Juvenile: scaly markings on back. Downy plumage like Skylark's. Voice: soft, trilling, soothing song, sung while flying, composed of short, descending, serial 'lū-lū-lū-lū-lū-', 'toolooeet toolooeet toolooeet' etc., call-note 'titlōō-eet'. Often heard also on clear nights. Nests in barren, thinly wooded regions and heaths. In winter seen in fields. Breeding season: IV—VI, usually 2 broods a year. Nest: in depression in ground, made of blades, roots and moss. Eggs: 3—5, whitish, thickly brown-speckled, incubated 13—15 days by ♀. Young fed 12—15 days by both parents in nest. Diet: insects, seeds. Partial migrant. Migration: II—III and IX—X.

3 **CRESTED LARK** *(Galerida cristata)* Size of Skylark. Fairly light; pointed crest, short tail with light brown outer feathers. ♂ = ♀. Juvenile: more spotted on back. Downy plumage: sparse, long, whitish. Voice: call-note 'twee-tee-too' also forms main motif of song, which is often heard on ground, sometimes in flight. Widespread nester on waste land, outskirts of towns, railway sidings, building sites, sports grounds, etc. Breeding season: IV—VI, 2 broods a year. Nest: untidy structure of blades of grass and rootlets in depression in ground. Eggs: 3—5, finely speckled, similar to those of Skylark. Care of brood and diet as for Skylark. Resident bird.

1

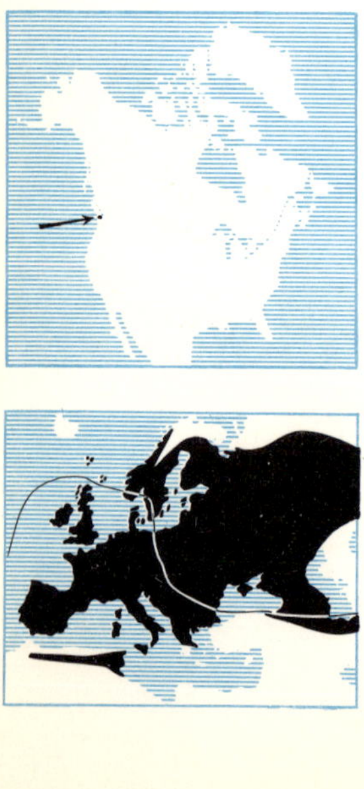

2

3

1
1
1 juv
2
2
2 juv
3
3
3 juv

Order: Passerines - Passeriformes **Family: Larks** - Alaudidae

1 **SHORT-TOED LARK** *(Calandrella cinerea)* Much smaller than Skylark. Unspotted, whitish underside, rather light, yellowish brown, vividly striped back. Short tail with white outer feathers. ♂ = ♀. Juvenile more spotted on back. Downy plumage light brown. Voice: call-note 'tchirrp', sings continually repeated 'tsit-siwee tsi-wichōō' motifs when flying. Undulating flight, close to ground. Nests in barren, sandy areas and dry or fallow fields in Mediterranean region. Breeding season: IV—VII, 2 broods a year. Nest: made of grass blades and fine parts of roots in hollow in ground. Eggs: 3—5, brownish yellow, finely speckled, incubated 13 days by ♀. Young fed by both parents. Diet: seeds, small beetles, other insects. Migrant. Migration: III—IV and X—XI.

2 **SHORE LARK - HORNED LARK** *(Eremophila alpestris)* Barely size of Skylark. Vivid black and yellow head markings, black band on crop. Back brown, speckled. ♂ has short, pointed 'ears', ♀ and juvenile less pronounced head markings. Ground-dweller, runs quite fast. Voice: metallic call-note 'twee-titi'; tuneful song sung mostly sitting, less often flying. Nests in tundra and on tundra-like mountain ridges; in winter seen on coast and occasionally inland (in stubble fields and on waste land). Breeding season: VI—VII, 1—2 broods a year. Nest: on ground, under tufts of grass, made of grass blades, padded with fluff and hair. Eggs: 4, like Skylark's, incubated about 12 days. Young fed 10—12 days by both parents. Diet: seeds of ground plants, insects. Migrant and partial migrant. Migration: III—V and X—XI.

Order: Passerines - Passeriformes **Family: Swallows** - Hirundinidae

3 **SWALLOW - BARN SWALLOW** *(Hirundo rustica)* Familiar form, with long, forked tail, russet throat and uniformly dark blue, gleaming back. Tail feathers have white spots in front of tips (cf. Red-rumped Swallow). ♂ distinguished from ♀ by longer 'prongs' of fork. Juvenile: back and throat duller. Downy plumage: a few long, grey down feathers. Voice: song hurried twittering with many notes and a few churring phrases; call-note 'tswit-tswit'. Common in farmland. Breeding season: V—VIII, 2—3 broods a year. Nest: dish made of lumps of clay stuck together with saliva, lined with straw and feathers, built indoors. Eggs: 4—5, white, with reddish brown spots, incubated 14—17 days by ♀. Young fed 20—22 days by both parents. Diet: flying insects. Migrant. Migration: IV and IX—X.

1

2

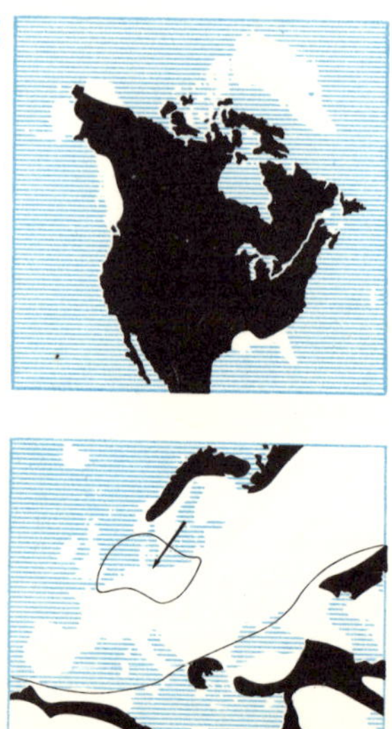

3

1
2
2 juv
3
3 juv
3

1 **HOUSE MARTIN** *(Delichon urbica)* Smaller than Swallow, with short-forked tail, pure white underside and conspicuous white rump. ♂ = ♀. Juvenile: brownish back, white-tipped inner secondaries. Downy plumage: sparse, long greyish white down feathers. Voice: call-note 'tsrrp' or 'tchirrip', song soft and twittering. Nests in farm settlements, frequently in colonies on outside of buildings, rarely on cliffs, concrete bridges, etc., away from human communities. Breeding season: V—VIII, 2 broods a year. Nest: closed hemisphere with small entrance, made of lumps of earth stuck together and lined with blades and feathers. Eggs: 4—5, pure white, incubated 13 days (and young fed 18—22 days) by both parents. Diet: insects. Migrant. Migration: IV—V and IX.

2 **CRAG MARTIN** *(Hirundo rupestris)* Size of House Martin. Brown back, straight-ended tail with row of white spots on posterior half. Dull white underside with no breast-band. ♂ = ♀. Juvenile: scaly back, fawn underside. Voice: 'tchrrri' or 'chit chit'. Breeding season: V—VII, 2 broods a year. Nests here and there in rocky ravines, in colonies. Nest: like Swallow's, in crack or niche in overhanging rock. Eggs: 4—5, white, with rust brown spots, incubated 14—15 days. Young fed 25—26 days by both parents. Diet: insects, including butterflies. Migrant. Migration: III and IX—X.

3 **SAND MARTIN - BANK SWALLOW** *(Riparia riparia)* Small and brown, without light rump. Brown breast-band, notched tail. ♂ = ♀. Juvenile: light-edged feathers, rust brown throat. Downy plumage pale grey. Voice: call-notes soft 'grrāā-grrāā' and harsh 'tchrrip' or 'tchrrr'. Nests in colonies in sand-pits and steep, sandy river banks. Breeding season: V—VII, 2 broods a year. Nest: at end of long, self-excavated, horizontal burrow, made of blades of grass, fibres and feathers. Eggs: 4—6, white, incubated 12—16 days (and young fed) by both parents. Young leave burrow at about 20 days. Diet: as for Swallow. Migrant. Migration: IV—V and VIII—IX.

1

2

3

1
2
3

Order: Passerines - Passeriformes **Family: Swallows** - Hirundinidae

1 **RED-RUMPED SWALLOW** *(Hirundo daurica)* Resembles Swallow in size and form, but has rusty yellow rump and nuchal stripe and buff underside. No white spots on tail feathers (cf. Swallow). ♂ = ♀. Juvenile: browner and duller. Voice: 'keet' when flying, warning call 'kēērr'. Nests in open country with rocks, also in towns. Breeding season: IV—VII, 2 broods a year. Nest: closed, with tubular entrance, made of clayey earth on rock or building. Eggs: 4—5, pure white, incubated about 14 days. Both parents feed young. Diet: small winged insects. Migrant. Migration: IV and IX—X.

Order: Passerines - Passeriformes **Family: Wagtails** - Motacillidae

2 **BLUE-HEADED WAGTAIL - YELLOW WAGTAIL** *(Motacilla flava)* About size of Sparrow, slim. ♀ has whitish, instead of yellow, throat. Juvenile: brown-spotted throat. Downy plumage: long, light brown down feathers. Walks daintily, flicks tail. Flies in long, flat curves. Voice: loud 'tsoueep' or 'tsirr'. Widespread nester in low-lying plains. Breeding season: V—VII, 1 brood a year. Nest: on ground. Eggs: 4—6, covered with greyish brown diffuse spots, incubated 13 days by ♀. Hatching period 11—12 days, both parents feed young. Diet: insects. Migrant. Migration: IV and VIII—IX. Several subspecies, easily distinguishable: A. Yellow Wagtail (*M. flava flavissima*, yellow crown, Great Britain), B. Ashy-headed Wagtail (*M. f. cinereocapilla*, grey head, white throat, Italy), C. Grey-headed Wagtail (*M. f. thunbergi*, dark grey crown, black ear-coverts, Scandinavia), D. Black-headed Wagtail (*M. f. feldegg*, black head, Balkans).

3 **GREY WAGTAIL** *(Motacilla cinerea)* Same size as White Wagtail, with sulphur yellow underside. Distinguished from Blue-headed Wagtail by ash grey back and distinctly longer tail. Habitat also different. ♂: in nuptial plumage black throat, in winter plumage white throat (like ♀). Juvenile: brown-grey back, brownish yellow underside. Downy plumage golden brown. Wags long tail, flies in flat curves. Voice: hard 'tzitzi' or 'zee-zee-zee' and 'siz-eet'. Nests by fast streams, sluices, water-mills, always to be found near water. Breeding season: IV—VII, 2 broods a year. Nest: loosely bound mass of blades of grass, rootlets and leaf fragments, lined with hair, usually in recess. Eggs: 4—6, yellowish, with russet diffuse spots and speckles, incubated 12—14 days (and young fed 12—13 days) by ♂ and ♀. Diet: insects. Partial migrant. Migration: III and X.

1

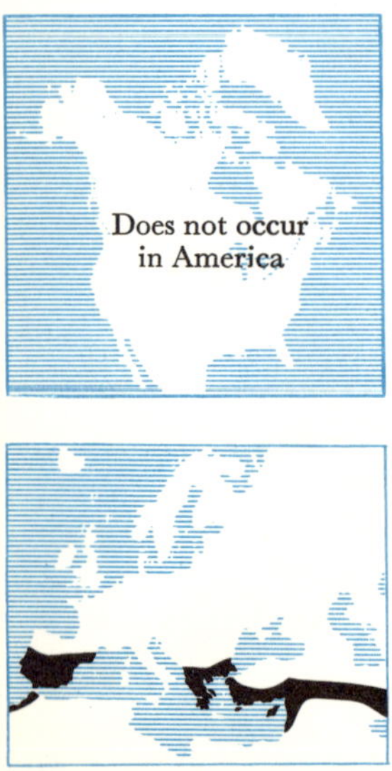

2

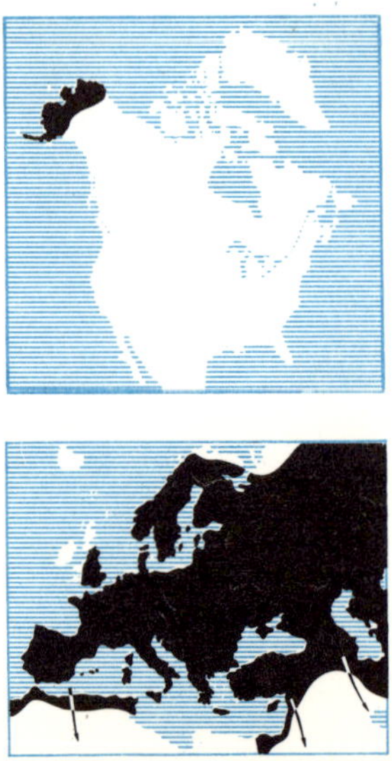

3

1
M. f. flava
M. f. flavissima
M. f. cinereocapilla
M. f. thunbergi
2 ♂
2 juv
M. f. feldegg
3 ♂ S
3 juv

1 **WHITE WAGTAIL** *(Motacilla alba)* Familiar species with vivid black and white markings. ♂ in winter plumage has white throat instead of black. ♀: black head markings intermingled with grey. Juvenile: brownish grey back, dark brown crop band. Downy plumage: smoky grey. Pied Wagtail *(1a — M. alba yarrellii)*, which nests in Britain, distinguished by black back. Voice: 'tzi-wirrp', when excited 'tchizzik', mixed with soft twittering in song. Flies in long curves. Nests mostly near water, often on buildings. Breeding season: IV—VIII, 2—3 broods a year. Nest: litter of blades of grass, leaves and rootlets, usually in recess. Eggs: 5—6, white, with thick dark grey spots, incubated 12—14 days by ♀. Young fed 14—15 days by both parents. Diet: insects. Partial migrant. Migration: III and X.

2 **WATER PIPIT** *(Anthus spinoletta)* Size of Sparrow. Slim, longish legs, dark feet. Nuptial plumage: unstriped, pink-tinted breast, grey-brown back, light eye stripe. Winter plumage: faint stripes down breast. Juvenile: darker spots. Downy plumage brownish grey. Flies hurriedly, in short curves. Voice: call-note 'tsip' or 'tsee-eep', song series of 'tsip-tsip-tsip-tsip' tones, sung during fluttering display flight. Nests in high mountain meadows. The more strongly spotted Rock Pipit *(A. s. petrosus)* inhabits rocky coasts of northern Europe. Breeding season: V—VII, 1—2 broods a year. Nest: made of grass blades and roots, hidden in crevice in ground, under clump of grass or bush. Eggs: 4—5, greyish white, thickly spotted, incubated 14—16 days by ♀. Young fed about 15 days by both parents. Diet: largely insects. Partial migrant. Migration: III—IV and VIII—IX.

3 **TAWNY PIPIT** *(Anthus campestris)* About size of Sparrow. Longish tail, long legs, conspicuous light eye stripe. Mostly struts about on ground, flies in flat curves. Differs from other pipits by fawn, almost unspotted plumage, appears lighter. ♂ = ♀. Juvenile: dark brown back with light-edged feathers, spotted crop. Downy plumage light brown. Voice: call-note sparrow-like 'tri-uc, tzeep', etc., song repeated 'chivee' calls, uttered while flying. Nests on dry, barren land and heaths with sand and bushes, in low-lying country. Breeding season: V—VII, 1 brood a year. Nest: on ground, hidden in heather or clump of grass. Eggs: 4—5, white, with a few brown spots, incubated 13—14 days by ♀. Young fed about 14 days by both parents. Diet: insects. Migrant. Migration: IV—V and VIII—IX.

1

2

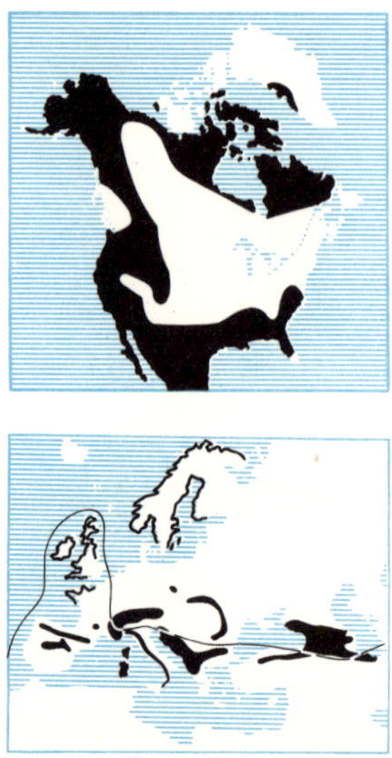

3

1 juv
1 ♂ S
1 ♂ S
1a ♂ S
2
2 S
2 W
3

1 **TREE PIPIT** *(Anthus trivialis)* Barely size of Sparrow. Back olive brown, with blackish stripes, breast yellowish, with stripes on sides of crop. Light eye stripe. ♂ = ♀. Juvenile: back more thickly striped. Downy plumage dark grey. Voice: call-note 'sip' or 'teeze', loud, tuneful song rendered during short flight, 'sipsipsipteezeteezeteeze' while rising, then an energetic trill and, while gliding with spread wings and tail, a loud 'seea-seea-seea'. Sits mostly in tree-tops, flies jerkily in short curves. Nests at margin of woods, in clearings and meadows with isolated trees, on heaths, etc. Breeding season: V—VII, 1 brood a year. Nest: on ground, hidden under clumps of grass or bush, made of blades and moss. Eggs: 5, mostly greyish white with brown speckles, incubated 12—13 days by ♀. Young fed about 12 days by both parents. Diet: insects. Migrant. Migration: IV—V and IX.

2 **MEADOW PIPIT** *(Anthus pratensis)* Smaller than similar Tree Pipit. More greyish green, with blurred dark stripes on back and smaller stripes on light breast. ♂ = ♀ = juvenile. Downy plumage brownish grey. Voice: call-note sharp 'tissip', different from note of Tree Pipit. Sings rather long phrase, 'tisp-tisp-tisp-til-til-til-tsi-tsi-tsi', etc., while flying. Usually runs about on ground, flies in 'hops'. Widespread bird in wet meadows and pastures, on moors and mountains; when migrating occurs in large parties in fields and beside ponds. Breeding season: IV—VI, 2 broods a year. Nest: on ground, made of blades of grass and moss, lined with hair. Eggs: 4—6, light grey with brown spots, incubated 13 days by ♀. Young fed 12—14 days by ♂ and ♀. Diet: insects, supplemented by seeds. Partial migrant. Migration: III—V and IX—XI.

3 **RED-THROATED PIPIT** *(Anthus cervinus)* Resembles Meadow Pipit, but back (especially rump) more thickly striped. In nuptial plumage, ♂ in particular distinguished by russet throat and breast. Juvenile = adult winter plumage. Downy plumage dark greyish brown. Voice: call-note sharp 'tsee-oz' (like Reed Bunting) or soft 'teu', quite different from Meadow Pipit. Nests in tundra belt, occasionally seen beside ponds and in meadows when migrating. Breeding season: VI—VII, 1 brood a year. Nest, eggs, and nest habits as for Meadow Pipit. Diet: insects, worms, molluscs, sometimes seeds. Migrant. Migration: IV—V and IX—X.

1

2

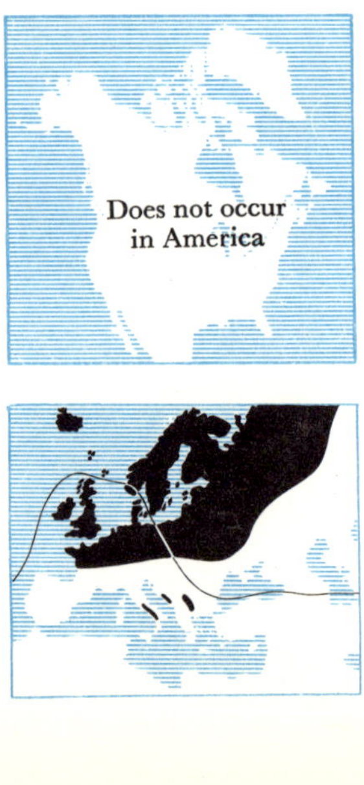

3

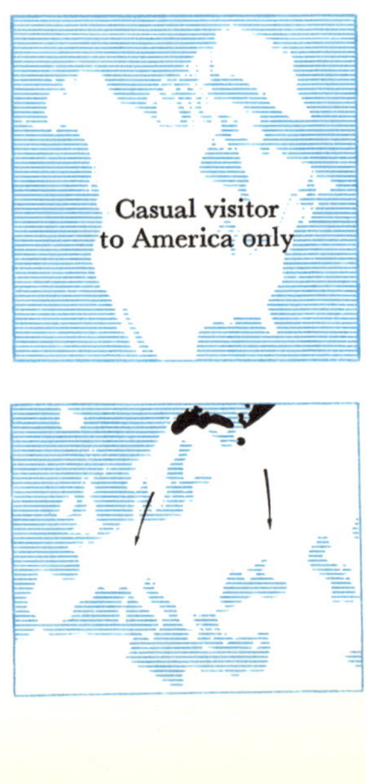

1
1
2
2 juv
2
3 S
3 W

Order: Passerines - Passeriformes **Family: Waxwings** - Bombycillidae

1 **WAXWING - BOHEMIAN WAXWING** *(Bombycilla garrulus)* Size of Starling. Reddish brown with pointed crest, white and yellow pattern on wings and tail. ♂ = ♀. Juvenile: chin and throat white instead of black. Voice: call-note a high trill, 'zhreee', mostly while flying. Flies like Starling. Nests in Scandinavian forests, in some years common winter bird in central Europe, usually in flocks. Breeding season: V—VI, 1 brood a year. Nest: in tree, made of twigs, lichen and grass blades, lined with hair and feathers. Eggs: 4—6, grey with a few dark brown spots, incubated about 14 days by ♀. Young fed by ♂ and ♀. Diet: in summer mainly insects, in autumn and winter almost solely berries. Migrant (winter bird). Migration: X—XI and III—IV.

Order: Passerines - Passeriformes **Family: Shrikes** - Laniidae

2 **GREAT GREY SHRIKE - NORTHERN SHRIKE** *(Lanius excubitor)* Size of Blackbird. Vivid black and light grey, strong, hook-tipped beak, wide black eye stripe, greyish white forehead (cf. Lesser Grey Shrike). ♂ = ♀. Juvenile: brownish grey back, ripple marks on breast and sides. Voice: warns intruders off nesting site by loud chattering; during courtship utters shrill 'trüü, kreerr' or 'wick' calls. Infrequent nester in open country with groups of trees. Often sits at top of tree, looking out for prey. Flies in flat curves, sometimes hovers. Breeding season: IV—VI, 1 brood a year. Nest: in tree, made of stalks, blades of grass and moss. Eggs: 5—6, white, with grey and brown spots, incubated about 15 days by ♀. Young fed by ♂ and ♀, 20 days in nest and about 2 weeks after fledging. Diet: insects, small vertebrates. Partial migrant and winter bird.

3 **LESSER GREY SHRIKE** *(Lanius minor)* Resembles Great Grey Shrike, but smaller, with black forehead and pink-tinted underside. ♂ = ♀. Juvenile: grey forehead, brownish grey back with ripple marks. Voice: chattering warning call, tuneful 'kviell' and muted, twittering mimetic song. Rare nester in lowland meadows and fields. Breeding season: V—VI, 1 brood a year. Nest: high in deciduous tree, made of brushwood and roots, mixed and decorated with green parts of plants. Eggs: 4—5, whitish, with brown spots, incubated about 15 days by ♀. Young fed by ♂ and ♀, about 2 weeks in nest and some time after while flying. Diet: insects. Migrant. Migration: V and VIII.

1

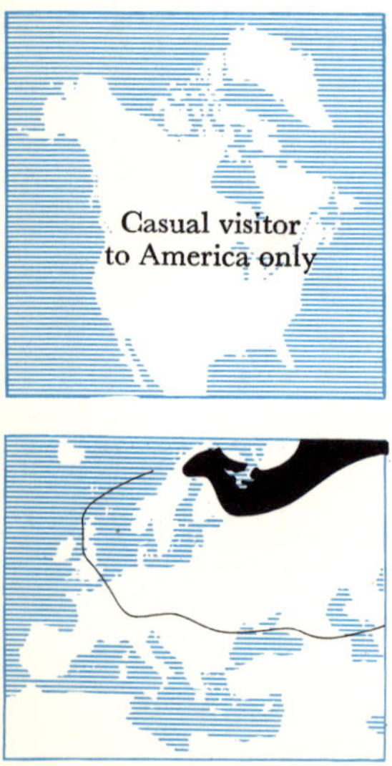

2

3

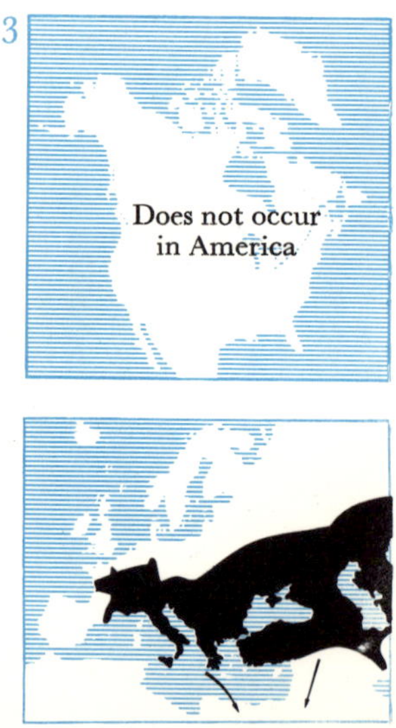

1
2
2 form with
2 specula
2 juv form
with 1 speculum
3
3
3 juv

Order: Passerines - Passeriformes **Family: Shrikes** - Laniidae

1 **RED-BACKED SHRIKE** *(Lanius collurio)* Size of large Sparrow. Large head, strong, hooked beak, wags tail. ♂: light grey head. ♀: rust brown head, ripple marks on breast and sides. Juvenile: like ♀, but ripple marks also on back. Downy plumage not developed. Voice: hard, raucous warning note 'chack, chee-uk' or serial 'gragragra', nestlings screech urgent 'keeuk'. Nests in open country with bushes, on thorn hedges, railway embankments, margin of woods, etc. Breeding season: V—VII, 1 brood a year. Nest: low in thicket, made of blades of grass, roots and moss. Eggs: 4—6, whitish, with rust brown spots which often form ring round blunt end, incubated 2 weeks by ♀. Young fed by ♂ and ♀, 2 weeks in nest and about 3 weeks after, while flying. Diet: insects, rarely small vertebrates and berries. Migrant. Migration: V and VIII—IX.

2 **WOODCHAT SHRIKE** *(Lanius senator)* Slightly larger than Red-backed Shrike, appears darker than other shrikes. Red-brown nape, white scapulars and rump. ♀: back greyish brown instead of black. Juvenile: greyish brown and white, cross-banded back. Voice: 'kiwick' calls, magpie-like chattering, sometimes harsh mimetic song. Sporadic nester in open lowlands and warm hilly country. Breeding season: V—VII, 1 brood a year. Nest: like Lesser Grey Shrike's, often on fruit-tree. Eggs: 5—6, white, with ring of brown spots at blunt end, incubated 15 days by ♀. Hatching period about 20 days. Diet: insects. Migrant. Migration: IV—V and VIII—IX.

Order: Passerines - Passeriformes **Family: Dippers** - Cinclidae

3 **DIPPER** *(Cinclus cinclus)* Size of Starling, with short, erect tail. Brown, throat and breast brilliant white. ♂ = ♀. Juvenile: slate grey back, white underside with dark cross-streaks. Downy plumage long, thick, dark grey. Voice: sharp 'zit' or 'cling', song low and twittering. Nests beside swift mountain streams with large boulders. Flight: whirring, near surface of water. Dives, swims and runs under water. Breeding season: IV—VII, 2 broods a year. Nest: mostly large, closed except for round entrance, built of moss and plant parts in hole in bank or ground, under bridge or weir. Eggs: 4—6, pure white, incubated 15—17 days (and young fed about 20 days) by ♂ and ♀. Diet: aquatic insects, freshwater shrimps, water snails. Resident and nomadic bird.

1

2

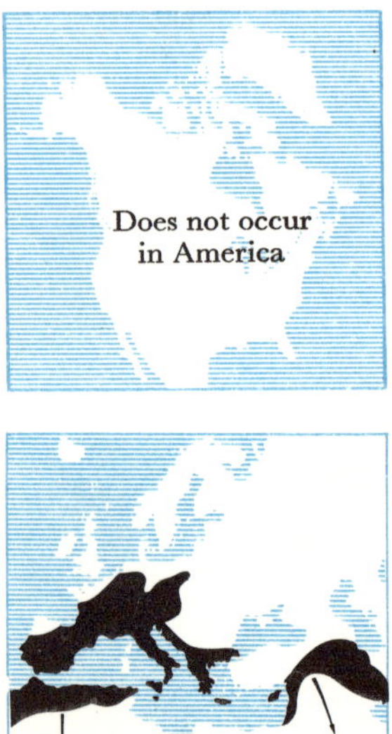

3

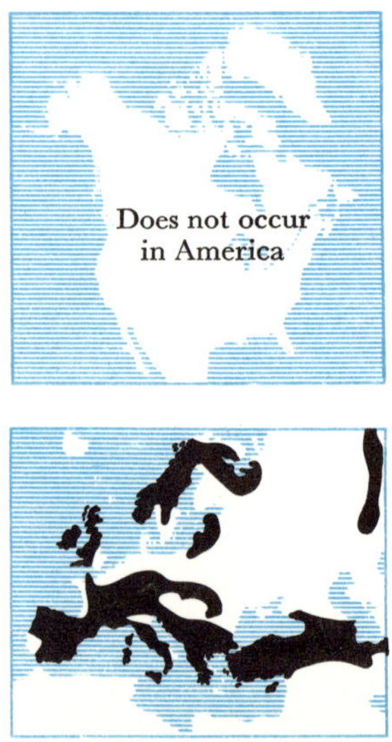

1 ♂
1 ♀
2 ♂
2 juv
3 juv
3
3

Order: Passerines - Passeriformes **Family: Wrens** - Troglodytidae

1 **WREN - WINTER WREN** *(Troglodytes troglodytes)* Tiny, with dark cross-bands and short, erect tail. ♂ = ♀ = juvenile. Downy plumage: sparse, blackish brown down featers. Voice: call-note 'tic-tic', warning note churring 'stirrrrrrup', song warbling phrase with loud trill. Always stays close to ground, flies whirringly for short distances only, scurries about in densest thickets. Common nester in woods with abundant undergrowth. Breeding season: V—VII, 2 broods a year. Nest: spherical, closed except for entry hole, made of moss, ferns, dry leaves, lined with feathers. Eggs: 5—7, white, with russet spots, incubated about 15 days by ♀. Nestlings fed 15—17 days by ♂ and ♀. Diet: insects. Partial migrant. Migration: III—IV and IX—X.

Order: Passerines - Passeriformes **Family: Accentors** - Prunellidae

2 **HEDGE SPARROW - DUNNOCK** *(Prunella modularis)* Smaller than Sparrow. Lead grey head and breast, striped brown back. ♂ = ♀. Juvenile: back rusty yellow, underside cream, with dark spots. Downy plumage: long, black down feathers. Voice: thin, repeated 'tseep' calls, song simple series of low, piping tones. Common forest bird, especially among young pines, occurs in mountains up to stunted timber level. Breeding season: IV—VII, 2 broods a year. Nest: near ground, hidden in thicket, made of few stalks and large amount of green moss, softly lined cup usually encrusted with red spore cases of moss. Eggs: 4—5, plain greenish blue, incubated 12—14 days, mainly by ♀. Young fed 13—14 days by ♂ and ♀. Diet: mainly insects, also small seeds, especially in winter. Partial migrant. Migration: III—IV and IX—X.

3 **ALPINE ACCENTOR** *(Prunella collaris)* Size of Sparrow. Front of body lead grey. White, black-speckled throat, rust brown-spotted sides. Rather short tail. ♂ = ♀. Juvenile: grey, unspotted throat, rest of underside rusty yellow with marked brown stripes. Downy plumage dark grey. Voice: lark-like call-note 'tchirp-rip', and sparrow-like 'churrg'; song similar to Skylark's, sung mostly on ground. Mountain bird, nests on rocky ground above forest belt. Breeding season: V—VII, 2 broods a year. Nest: made of moss, blades of grass and roots, softly lined cup often encrusted with red spore cases of moss. Eggs: 4—5, plain bluish green, incubated about 15 days (and young fed) by ♂ and ♀. Diet: insects, spiders, snails and slugs, seeds. Partial migrant. Migration: IV and X.

1

2

3

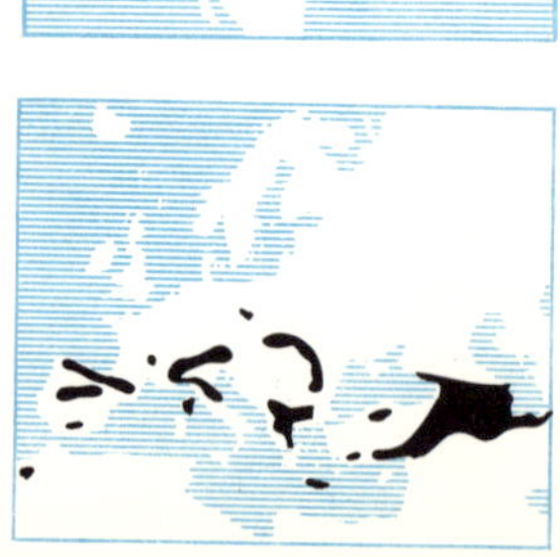

1 juv
1
2
2 juv
3 juv
3

1 **FAN-TAILED WARBLER** *(Cisticola juncidis)* Small reed warbler, hardly larger than Goldcrest. Brown, dark-striped back, rust brown rump. Further differentiated from similar Sedge Warbler by dark markings on crown and absence of light eye stripe. Short, rounded tail with white-tipped feathers. ♂ = ♀ = juvenile. Voice: song, 'cheek-tew-dzeep-dzeep', is sung during flight, in rhythm with wing strokes. Nests in lowlands, in damp spots with marsh vegetation and bushes, also in corn- and rice-fields. Retiring bird. Breeding season: IV—VII, 2 broods a year. Nest: pouch-shaped, with entry hole at top, made of dry leaves and plant fibres, close to ground, cleverly woven between thick clumps of plants. Eggs: 4—6, usually white with russet spots. Diet: insects. Resident bird.

2 **CETTI'S WARBLER** *(Cettia cetti)* Almost size of Sparrow. Plain, dark russet back, greyish white underside, widely rounded tail. ♂ = ♀ = juvenile. Downy plumage almost black. Retiring bird, but mostly given away by voice: chattering warning call, loud 'chik' and 'chi-wik' calls and trumpeting song, 'weechōō-cheweechōō-cheweechōō,weechōō'. Lives and nests hidden in waterside reeds, often beside ditches. Breeding season: IV—VI, 2 broods a year. Nest: cup-shaped, made of stems, grass blades and moss in densest thicket, close to ground. Eggs: usually 4, brick-red, unspotted, incubated 13 days by ♀. Young fed also by ♀, 14 days in nest and 14 more days while flying. Diet: insects. Resident and nomadic bird.

3 **MOUSTACHED WARBLER** *(Acrocephalus melanopogon)* Small reed warbler. Distinguished from similar Sedge Warbler by almost black crown, wide, pure white eye stripe and rump (same shade as back). ♂ = ♀ = juvenile. Voice: warning note 'churr', like Sedge Warbler's, song also similar, but softer and often introduced by four notes reminiscent of Nightingale's song. Infrequent nester by shallow ponds, lakes and in marshes, often nests at edge of large reed-beds. Breeding season: IV—VI, 2 broods a year. Nest: cup-shaped, made of reed-grass, rush and sedge leaves, attached to reed blades just above water. Eggs: 4—5, whitish, brown-speckled. Both ♂ and ♀ incubate eggs and feed young. Diet: insects. Partial migrant. Migration: IV and IX.

1

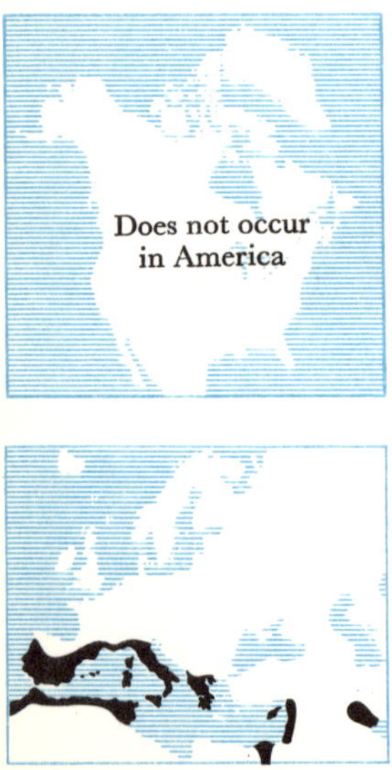

2

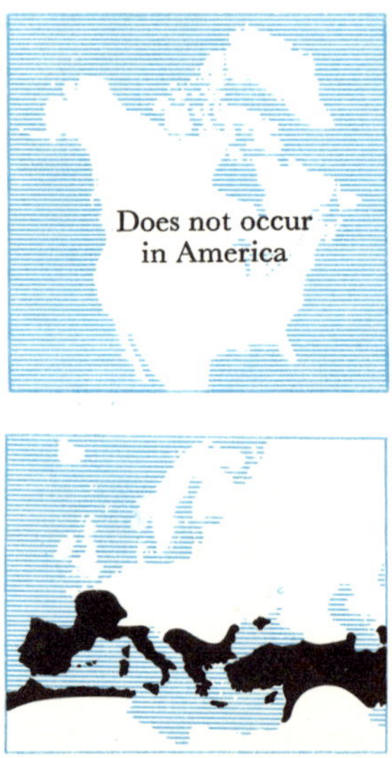

3

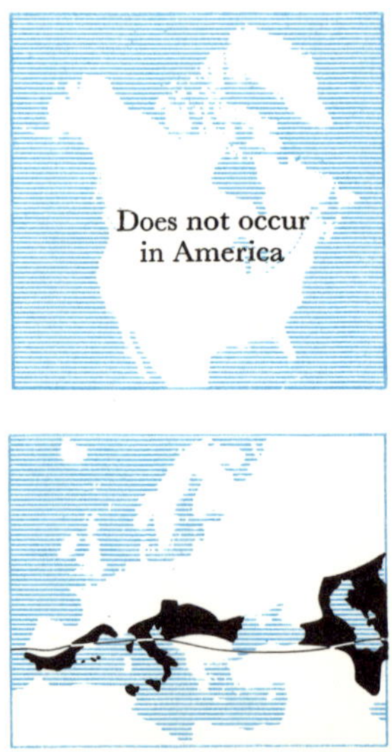

1
1
2
3 juv
3

1 **SAVI'S WARBLER** *(Locustella luscinioides)* Smaller than Sparrow. Plain reddish brown, no markings. Underside rusty white. Wide tail rounded off in steps. ♂ = ♀ = juvenile. Downy plumage: sparse reddish brown down feathers. Voice: call-note, sharp 'tzwik', heard only at close quarters, song monotonous 'urrrrrrr', like winding angler's reel (deeper than Grasshopper Warbler's song). Retiring bird, lives in swampy regions with spreading sedge- and reed-beds, often dotted with willow-bushes. Frequently climbs reeds while singing. Breeding season: V—VII, 2 broods a year. Nest: fairly large, made of rush blades and leaves, in thickest reed-grass over water. Eggs: 4—5, whitish, with reddish brown speckles, incubated 12 days (and young fed 12—14 days) by ♂ and ♀. Diet: insects. Migrant. Migration: IV and VIII—IX.

2 **RIVER WARBLER** *(Locustella fluviatilis)* Almost size of Sparrow. Umber, whitish underside with indistinctly striped breast. Wide tail rounded off in steps. ♂ = ♀ = juvenile. Voice: song persistent, sharp, buzzing 'derrr derr, derr derr, derr derr . . .', often resembling first part of Yellowhammer's song in pitch and rhythm. Also sings regularly at night. Nests in riverside woods with dense undergrowth (stinging-nettles, bushes, brambles, etc.), sometimes in overgrown clearing or marshy meadow with willows. Breeding season: V—VII, 1—2 broods a year. Nest: cup-shaped, made of blades of grass, reed-grass leaves and moss in densest thicket, close to ground. Eggs: 4—5, white, russet-speckled, incubated about 13 days. Diet: insects. Migrant. Migration: V and VIII.

3 **GRASSHOPPER WARBLER** *(Locustella naevia)* Much smaller than Sparrow. Back olive brown, thickly streaked, underside plain yellowish white. ♂ = ♀. Juvenile: a few stripes on crop. Downy plumage dark grey, long and thick. Voice: song a long metallic 'seeerrrrr' closely resembling monotonous chirping of Green Grasshopper and heard mainly at night. Nests in damp meadows with willow-bushes, beside swampy ponds, sometimes in rape- and clover-fields or in clearings. Breeding season: V—VII, 1—2 broods a year. Nest: deep cup made of grass leaves, on ground, in tangled vegetation. Eggs: 5—6, pink, russet-speckled, incubated 14 days (and nestlings fed 10—13 days) by ♂ and ♀. Diet: insects. Migrant. Migration: IV—V and VIII—IX.

1

2

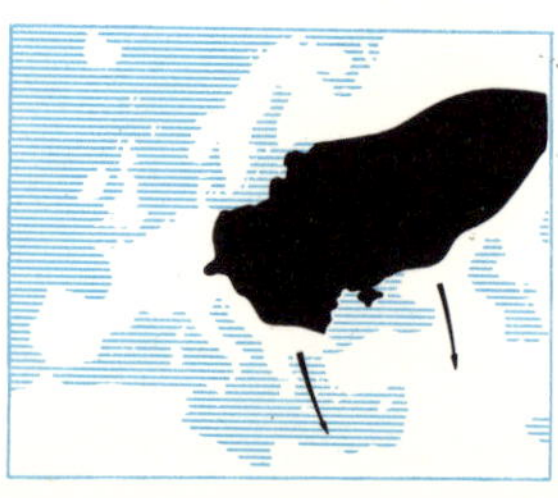

3

1
2
3
3
3 juv

1 **SEDGE WARBLER** *(Acrocephalus schoenobaenus)* Much smaller than Sparrow. Olive brown back thickly striped, rust brown rump unstriped. Dark crown, vivid black and white eye stripe, white throat. ♂ = ♀. Juvenile: brighter colouring. Downy plumage not developed. Voice: warning notes 'tuk' or snarling 'churr'. Song hurriedly rendered, varied phrases with long trills and 'wheet' calls, also sung at night. ♂♂ often sing during short display flight. Nests in marshy country with sedge and bushes, beside ponds, rarely in corn-fields. Breeding season: V—VII, usually only 1 brood a year. Nest: strong cup made of grass and moss, low down among reeds or sedge or in thicket. Eggs: 4—6, thickly covered with rust brown speckles, incubated 12—13 days (and nestlings fed 12—16 days) by ♂ and ♀. Diet: insects. Migrant. Migration: IV and IX—X.

2 **AQUATIC WARBLER** *(Acrocephalus paludicola)* Resembles Sedge Warbler in size and markings, but has yellow crown stripe as well as light eye stripes and stripes on back extend to rump and sides. ♂ = ♀ = juvenile. Voice: like Sedge Warbler's, but song less varied. Often sings during short display flight. Runs about on ground. Nests in marshes with large sedge meadows, seen beside ponds when migrating. Breeding season: V—VI, 2 broods a year possible. Nest and eggs: as for Sedge Warbler, little known of nesting habits. Diet: insects. Migrant. Migration: IV—V and VIII—IX.

3 **GREAT REED WARBLER** *(Acrocephalus arundinaceus)* Almost Starling size, largest reed warbler. Plain brown, with indistinct light eye stripe, ♂ = ♀ = juvenile. Downy plumage absent. Voice: highly distinctive, loud, squawking song, 'gurk-gurk-gurk-keek-keek-keek-karra-karra-keek', resolute, croaking warning note. Often sings at night. Widespread nester in reed-banks in regions with ponds and lakes. Breeding season: V—VII, usually only 1 brood a year. Nest: deep cup made of grass leaves and reed fibres, cleverly attached to reed blades over water. Eggs: 5—6, bluish white, with green to brown spots, incubated 13—15 days (and nestlings fed about 12 days) by ♂ and ♀. Diet: insects. Migrant. Migration: V and VIII—IX.

1

2

3

1 juv
1
1
2
3

1 **REED WARBLER** *(Acrocephalus scirpaceus)* Resembles Great Reed Warbler, but smaller even than Sparrow (also cf. Marsh Warbler). ♂ = ♀ = juvenile. Downy plumage not developed. Alarm notes harsh 'churr' and 'karrr', also hoarse 'gurk'. Song not loud, keeps strict time, with triple motifs, 'hirruc-chirruc-chirruc-teer-tuk-tuk-tuk-jag-jag-churr-churr-churr', etc., also sung at night. Common nester in reed-beds beside ponds, lakes and in marshes, sometimes in willow-beds. Breeding season: V—VII, 1—2 broods a year. Nest: like Great Reed Warbler's, but smaller, usually attached to reed blades standing in water. Eggs: 4—5, white, with grey-brown speckles, incubated 11—12 days (and young fed 11—14 days) by ♂ and ♀. Diet: insects. Migrant. Migration: V and IX—X.

2 **MARSH WARBLER** *(Acrocephalus palustris)* In field observations, definitely distinguishable from Reed Warbler only by voice. Back also more olive brown. ♂ = ♀ = juvenile. Downy plumage absent. Voice: sharp 'tuc' and 'tchuc'. Varied song reminiscent of Icterine Warbler's, usually contains introductory 'tic-tic-tic' calls, warbling 'wheet-wheet-wheet', tuneful rolls and a number of mimetic motifs, also sung at night. Nests in damp meadows, beside nettle-infested ditches, on river banks, in thickets and often in rye-fields. Breeding season: V—VII, 1 brood a year. Nest: hidden in tangled vegetation, made of dry grass, rootlets and a few hairs. Eggs: 4—5, bluish white, with large brown spots, incubated 12—13 days (and young fed 11—13 days) by ♂ and ♀. Diet: insects, supplemented by berries. Migrant. Migration: V and IX.

3 **ICTERINE WARBLER** *(Hippolais icterina)* Smaller than Sparrow. Grey-green back, yellow underside, short yellow eye stripe, reddish brown beak. Distinguished from similar Melodious Warbler by yellow area on wings and bluish grey legs. ♂ = ♀ = juvenile. Downy plumage absent. Voice: call-note 'deederoid'. Song long, sharp and twittering, usually with triple motifs and often including typical 'deederoid'. Nests in open deciduous woods and in villages, keeps mostly to top of deciduous trees. Nest: strong cup made of plant and animal (spiders' webs) matter, neatly smoothed and generally lined outside with slivers of birch bark, built not too far above ground. Eggs: 5, rose pink, with a few black spots, incubated 13 days (and young fed 13 days) by ♂ and ♀. Diet: insects, supplemented by berries and fruit. Migrant. Migration: V and VIII.

1

2

3

1
2
3

1 **OLIVACEOUS WARBLER** *(Hippolais pallida)* Size and form like Icterine Warbler. Light umber back, whitish underside (without yellow), light eye stripe. Rather short wings, longer beak than Icterine Warbler's. No light area on wings. ♂ = ♀ = juvenile. Voice: call-note short 'tec', song less sharp and rich than Icterine Warbler's, more like song of Sedge Warbler. Nests in gardens and parks, hedges and light riverside woods, frequents thickets. Breeding season: V—VI, 1 brood a year. Nest: like Icterine Warbler's, in hedge, seldom in tree. Eggs: 4—5, reddish white, with black speckles, incubated 14—15 days. Hatching period about 15 days. Diet: insects. Migrant. Migration: IV—V and VIII.

2 **MELODIOUS WARBLER** *(Hippolais polyglotta)* Closely resembles Icterine Warbler, but, on closer observation, has shorter wings, usually lacks the pale wing patch and has brownish legs. ♂ = ♀ = juvenile. Downy plumage not developed. Voice: softer and more melodious than Icterine Warbler's, a quick twittering with frequently recurring chirruping sounds and mimetic motifs from other bird-songs; sometimes a short display flight. Nests in western Europe, habitat similar to Icterine Warbler's. Breeding season: V—VII, mostly 1 brood a year. Nest: like Icterine Warbler's, but smaller and less strongly built, usually in thicket. Eggs: 4—5, reddish, white with black spots, incubated 13 days by ♀. Nestlings fed 12—13 days by ♂ and ♀. Diet: insects. Migrant. Migration: IV—V and VIII—IX.

3 **GARDEN WARBLER** *(Sylvia borin)* Almost size of Sparrow. Plain greyish brown, with lighter underside. No distinctive markings. Feet lead grey. ♂ = ♀. Juvenile: back more rust brown, underside more yellowish brown, without white throat. Downy plumage not developed. Voice: warning note 'whet-whet' or hard 'check'. Song, rendered during rapid flight, long, pleasing, prattling verse without great variation in pitch. Common nester in parks, forests with plenty of undergrowth, bushes beside ponds, etc. Breeding season: V—VII, 1 brood a year. Nest: quite large, loosely made of blades of grass, rootlets, moss and stalks, usually low down in bushes. Eggs: 5, yellowish white, with brown spots, incubated 13—14 days (and young fed 11—12 days) by ♂ and ♀. Diet: insects, berries, fruit. Migrant. Migration: V and VIII—IX.

1

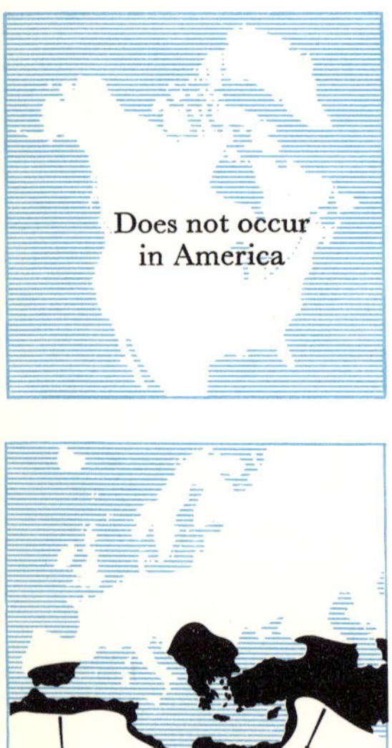

2

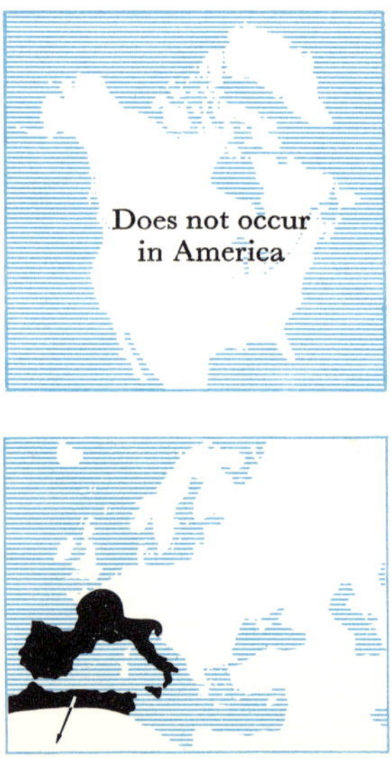

3

1
2
3 juv
3

1 **BLACKCAP** *(Sylvia atricapilla)* Almost size of Sparrow. ♂: grey-brown back, sharply defined, glossy black cap reaching to eyes, sides of head grey. ♀: russet cap. Juvenile: similar to ♀. Downy plumage absent. Stays mainly in dense thicket, near ground. Voice: frequent, sharp, warning 'tacc'. Complete song has twittering prelude, followed by rich-toned, fluty, short finish song. Nests in open woods with undergrowth, parks, mountain forests. Breeding season: V—VII, 2 broods a year. Nest: small cup, loosely built, made of blades of grass, roots and hair. Eggs: 5—6, white, with reddish brown spots, incubated 12—14 days (and nestlings fed 10—13 days) by ♂ and ♀. Diet: insects, berries, fruit. Partial migrant. Migration: IV and IX—X.

2 **SARDINIAN WARBLER** *(Sylvia melanocephala)* Generally resembles Blackcap, but has grey back. Glossy black cap (in ♀ ash grey) extends to cheeks and is sharply separated there from pure white throat. Crimson orbital ring. Tail rounded, outer feathers white. Juvenile = ♀. Downy plumage not developed. Voice: alarm notes rattling 'trrrrr' or 'cha-cha-cha-cha', song partly resembles Whitethroat's, but is longer and sweeter. Also sings during short display flight. Mediterranean species, inhabits dry country with dense thickets, open woods with undergrowth and vineyards. Breeding season: III—VI, 2 broods a year. Nest: made of dry stalks and grass. Eggs: 3—5, yellowish white, with russet speckles, incubated 13—14 days (and young fed 11—12 days) by ♂ and ♀. Diet: insects, spiders, berries, fruit. Resident and nomadic bird.

3 **ORPHEAN WARBLER** *(Sylvia hortensis)* Large warbler with dull black cap extending to below white eyes. Whitish throat, slate grey back, white outer tail feathers. ♀ and juvenile: back browner, cap greyer. Voice: rattling alarm note, call-note 'tac', etc., song loud and reedy, with repeated 'chiwirroo' and 'tittiwoo' motifs. Nests in open Mediterranean woods rich in undergrowth, olive groves, orchards. Breeding season: V—VII, 1 brood a year. Nest: mostly low down in thicket, loosely made of stalks, grass blades and plant fluff. Eggs: 5, white, with a few brownish black spots, incubated 12—13 days by ♀. Nestlings fed about 12 days by ♂ and ♀. Diet: insects, berries, fruit. Migrant. Migration: IV—V and VIII—X.

1

2

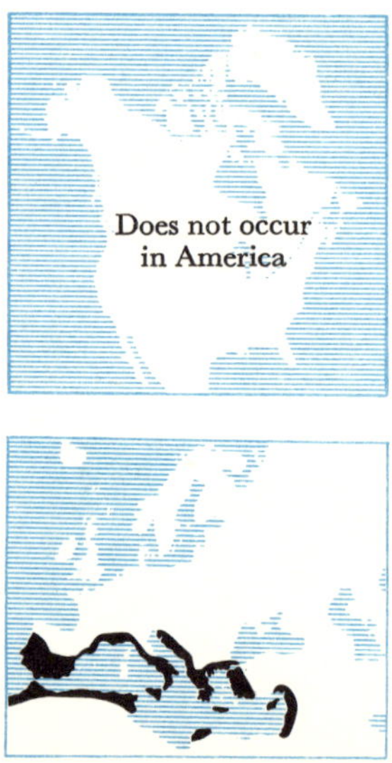

3

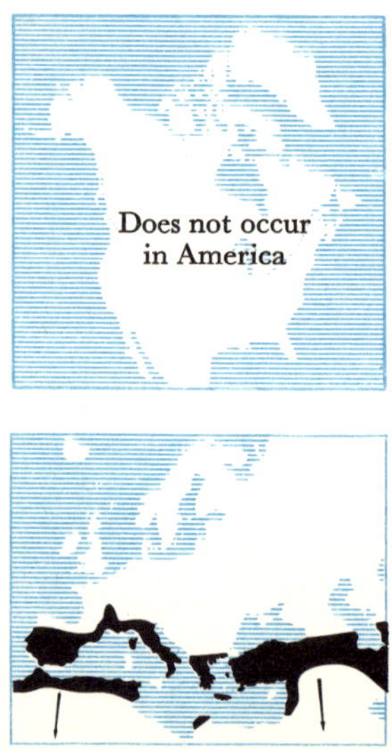

1 ♀
1 ♂
2 ♀
2 ♂
2 ♂
3 ♀
3 ♂

1 **LESSER WHITETHROAT** *(Sylvia curruca)* Small, with dark grey back, dark brown wings, blackish ear-coverts, white throat, white outer tail feathers. ♂ = ♀ = juvenile. No downy plumage. Voice: call-note hard 'tacc'; song in two parts, rough, hurried twittering prelude and warbling 'chikka-chikka-chikka-chikka-chikk' as main song. Nests in hedges, bushes, gardens, young conifer woods. Breeding season: V—VII, 1 brood a year. Nest: loosely made of dry stalks and grass blades, usually low down in dense thicket. Eggs: 5, white, with a few reddish brown spots, incubated 11—13 days (and nestlings fed also 11—13 days) by ♂ and ♀. Diet: insects, berries. Migrant. Migration: IV and IX—X.

2 **WHITETHROAT** *(Sylvia communis)* Almost size of Sparrow. Ash grey head, pure white throat, upper surface of wings bright russet, outer tail feathers white. ♀: head brownish grey (= juvenile). No downy plumage. Voice: a frequent muted 'wheet-wheet-whit-whit-whit' and sharp warning note 'tacc'. Song rough twittering, often with loud finish phrase sounding like 'cheechiwee-cheechiweechōō-chiwichōō'. Usually sings from height (top of bush, telephone wires), sometimes in short display flight. Inhabits thickets, e.g. sloe, rose and bramble hedges, willow-bushes, young conifer plantations, margins of woods. Breeding season: V—VII, 2 broods a year. Nest: low down in thicket, made of dry stems, grass blades, plant fluff, wool. Eggs: 4—6, greenish white, thickly brown-speckled, incubated 12—13 days (and young fed 13 days) by ♂ and ♀. Diet: insects, berries. Migrant. Migration: IV—V and IX.

3 **DARTFORD WARBLER** *(Sylvia undata)* Small, dark, with dark wine red underside and long, erect, usually fanned tail. ♂ = ♀. Juvenile: altogether browner. Voice: warning call metallic 'tchir-r-tuc-tuc', song resembles Whitethroat's, but sweeter, often rendered from height (e.g. telephone wires). Nests in dry country with large quantity of thorn-bushes, thickets, gorse and heather. Breeding season: IV—VI, 2 broods a year. Nest: usually low down in thorn-thicket, made of stems, grass blades and moss. Eggs: 3—4, whitish, brown-speckled, incubated about 12 days, mostly only by ♀. Nestlings fed 11—13 days by ♂ and ♀. Diet: insects, spiders. Resident and nomadic bird.

1

2

3

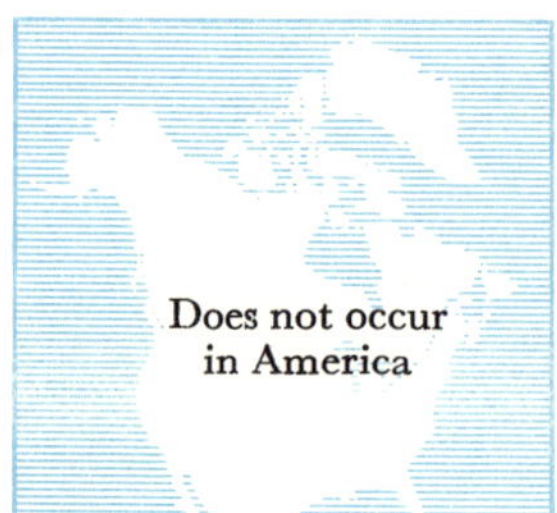

1
2
2 juv
3 ♂
3 juv

1 **SUBALPINE WARBLER** *(Sylvia cantillans)* Small warbler. Head and back slate grey, throat and breast russet. Distinctive white moustachial stripe. ♀ and juvenile: in general duller, underside yellower. Voice: song resembles Whitethroat's, but sweeter, often with short display flight. Mediterranean species, inhabits dry, sunny slopes with plenty of bushes and a few oaks. Breeding season: IV—VI, 2 broods a year. Nest: low down in thicket, resembles other warbler nests. Eggs: 4, whitish, thickly covered with brown spots, incubated 11—12 days, mainly by ♀. Young fed 12 days by ♂ and ♀. Diet: insects. Migrant. Migration: III—IV and VIII.

2 **BARRED WARBLER** *(Sylvia nisoria)* Size of Sparrow, robust. Ash grey, banded sides and breast, bright yellow eyes. ♀: back browner, cross-bands fainter. Juvenile: like ♀, with practically no markings. Voice: conspicuous snarling 'tcharrr', also 'tchack' calls. Song somewhat resembles Garden Warbler's, but is shorter and often interspersed with 'tcharrr' notes. Inhabits mountainsides, riverside meadows and fields with thickets and hedges, especially thorn-bushes. Breeding season: V—VI, 1 brood a year. Nest: quite large, built low down in thorn-bush, made of stems, blades, roots and hair. Eggs: 5—6, whitish, with light grey speckles, incubated 15—16 days (and nestlings fed 14—16 days) by ♂ and ♀. Diet: insects, berries, fruit. Migrant. Migration: V and VIII.

3 **CHIFFCHAFF** *(Phylloscopus collybita)* Small leaf warbler. Dainty, with dark greyish green back. Underside lighter shade of same colour. Dark legs. Indistinct, yellow eye stripe. ♂ = ♀ = juvenile. Distinguished from Willow Warbler by colour of legs and quite different song. Voice: call-note soft, piping 'hweet', song regular staccato 'chiff, chaff, chiff, chaff . . .' Common bird in woods with plenty of undergrowth, in mountains to limit of forest belt. Breeding season: IV—VII, 1—2 broods a year. Nest: on ground or in dense vegetation or bush near ground, oven-shaped, made of dry blades of grass, leaves and moss, lined with feathers. Eggs: 6—7, white, with brown spots, incubated 14—15 days by ♀. Nesting period 12—14 days, usually brooded and fed only by ♀. Diet: insects. Partial migrant. Migration: III—IV and VIII—XI.

1

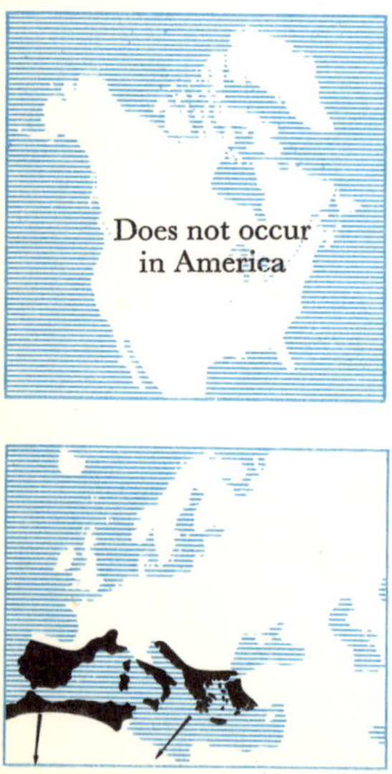

2

3

1 ♂
1 ♀
2
2
2 juv
3

1 **WOOD WARBLER** *(Phylloscopus sibilatrix)* Smaller than Sparrow, dainty. Green back. Very active, mainly frequents tree-tops. Light yellow throat and breast, white belly (distinct from other *Phylloscopus* species). Conspicuous wide yellow eye stripe. Pale legs. ♂ = ♀ = juvenile. Downy plumage light grey. Voice: song composed of series of 'stip' notes repeated at same pitch, with increasing rapidity, to 'stip, stip, stip, stip-stip-stip-stip - shreeeee'. Often sung during fluttering display flight, introduced by fluty 'piu-piu-piu'. Nests in mature deciduous woods, especially beech-woods. Breeding season: V—VI, 1 brood a year. Nest: on ground, spherical structure made of grass, side entrance. Eggs: 6—7, white, with brown spots, incubated 13 days by ♀. Young fed 13 days by ♂ and ♀. Diet: insects, berries. Migrant. Migration: IV—V and VIII—IX.

2 **WILLOW WARBLER** *(Phylloscopus trochilus)* Hard to distinguish from Chiffchaff in open, but usually yellower, with more distinct yellow eye stripe. Light-coloured legs. ♂ = ♀ = juvenile. Downy plumage greyish white. Very active, flutters adroitly through foliage. Voice: call note enquiring 'hooeet', song slowly descending, whistling verse vaguely resembling song of Chaffinch. Common nester in all types of woods, parks and often willow-bushes. Breeding season: IV—VI, 1 brood a year. Nest: on ground, in grass at edge of thicket, oven-shaped, made of dry grass and leaves, lined with feathers. Eggs: 6—7, white, with rust-coloured spots, incubated 13—15 days by ♀. Young fed 13—16 days by ♂ and ♀. Diet: insects. Migrant. Migration: IV and VIII—X.

3 **BONELLI'S WARBLER** *(Phylloscopus bonelli)* Greyish brown leaf warbler. Underside all white, rump yellowish green. (Wood Warbler also white-bellied, but has yellow throat and pronounced yellow eye stripe.) ♂ = ♀ = juvenile. Voice: call note usually disyllabic 'hou-eet', song brief warble consisting of several notes at same pitch. Very active bird, frequents open tree canopy. Nests in mixed woods with large proportion of conifers, especially on dry mountain-sides. Breeding season: V—VI, 1 brood a year. Nest: on ground, spherical, with side entrance, made of leaves, moss and blades of grass. Eggs: 5—6, white, thickly covered with red-brown spots, incubated 13—14 days by ♀. Hatching period 12 days, young fed by ♂ and ♀. Diet: insects, spiders. Migrant. Migration: IV—V and VIII—IX.

1

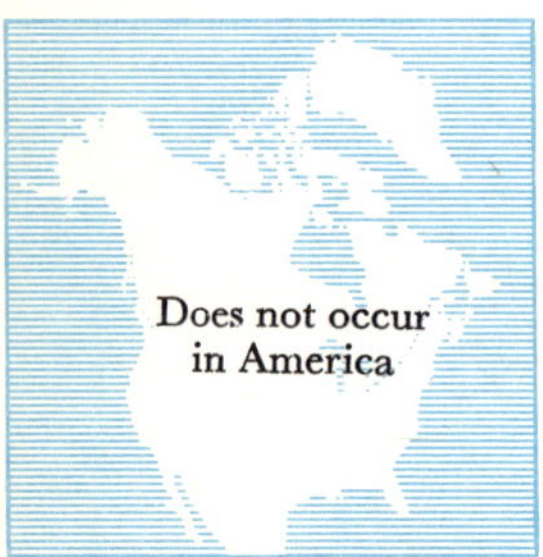

2

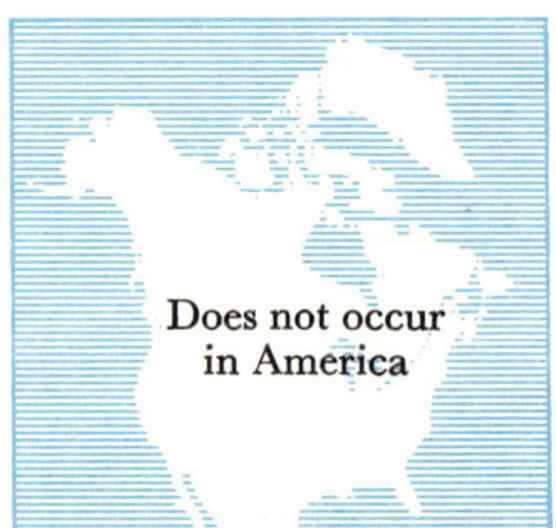

3

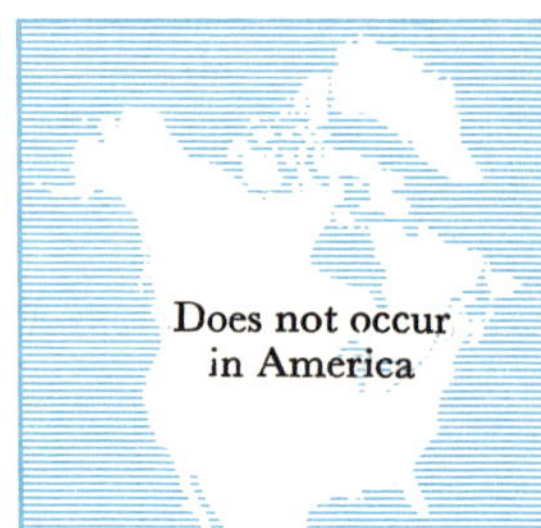

1
1
2
3

Order: Passerines - Passeriformes **Family: Warblers** - Sylviidae

1 **GREENISH WARBLER** *(Phylloscopus trochiloides)* Closely resembles Chiffchaff, but has narrow yellow wing band and eye stripe. ♂ = ♀ = juvenile. Best identified by voice: call note 'tsip' or 'tsoueep', similar to Blue-headed Wagtail's, song loud trill with twittering tones, rolling like Wren's song. Scattered nester in gardens, parks and woods. Breeding season: V—VI, 1 brood a year. Nest: on ground, among grass or in thicket, made of blades of grass, moss and dead leaves, oven-shaped. Eggs: 5—6, plain white, incubated 12—14 days by ♀. Young fed 12—15 days by ♂ and ♀. Diet: insects. Migrant. Migration: V—VI and IX.

Order: Passerines - Passeriformes **Family: Goldcrests** - Regulidae

2 **GOLDCREST** *(Regulus regulus)* Minute, very active bird. Olive green, with lighter underside, two white bars and one black bar on wings. Black-bordered crown is bright orange in ♂, yellow in ♀. Juvenile: plain crown. Downy plumage dark grey. Voice: call-note thin 'zeezeezee', song high, whispering, rising and falling series of notes. Common nester in conifer forests. Breeding season: V—VI, 2 broods a year. Nest: deep, thick-walled cup made of moss and lichen, attached to undersurface of fir twigs. Eggs: 8—10, whitish, with faded light grey markings, incubated about 16 days by ♀. Hatching period 15—16 days, young fed by ♂ and ♀. Diet: insects, spiders. Partial migrant. Migration: III—IV and IX—X.

3 **FIRECREST** *(Regulus ignicapillus)* Like Goldcrest, but more brightly coloured. Head markings supplemented by white eye stripe and black lores. Plumage on shoulders golden yellow. ♀ and juvenile: same differences as for Goldcrest. Voice: song resembles Goldcrest's, but does not change pitch, gathers volume slowly, sounds like 'zeezeezeezeezeezia'. Common nester in conifer forests. Breeding season: V—VII, 2 broods a year. Nest, eggs, care of brood and diet as for Goldcrest. Migrant. Migration: IV and X.

1

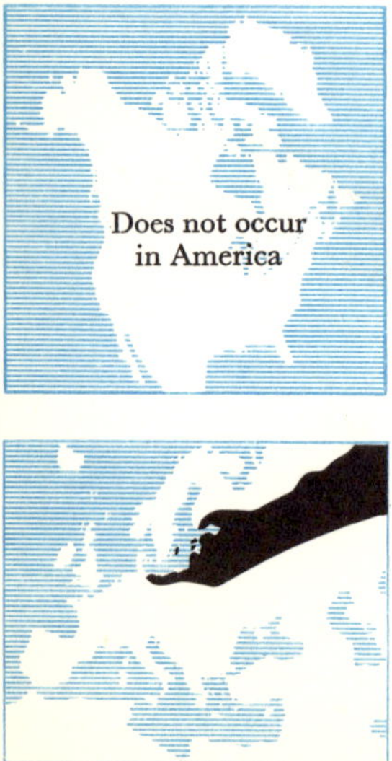

2

3

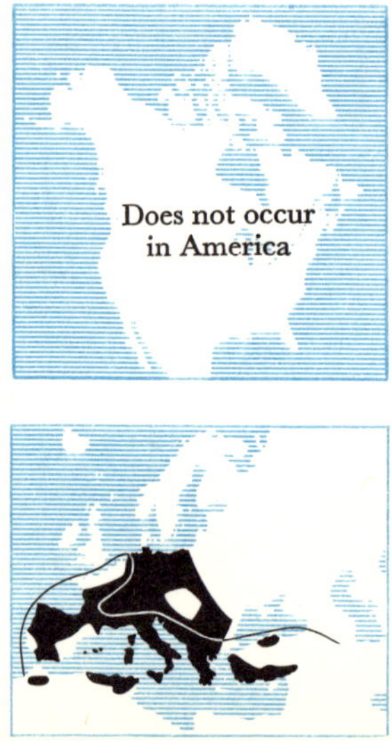

1
2 ♂
2 juv
2 ♀
3 ♀
3 ♂

1 **SPOTTED FLYCATCHER** *(Muscicapa striata)* About size of Sparrow. Back grey-brown, underside white, with thin dark lines down breast and over head. ♂ = ♀. Juvenile: light-edged feathers especially on back, appears spotted. Downy plumage dark grey. Mostly sits erect on end of dead twigs, telegraph wires, fences, etc., often twitches wings and tail and flies up a few yards to catch flying insects. Voice: call-note sharp 'tzee', warning call 'tuc-tuc'. Frequent nester in parks, gardens, at edge of woods, often in villages. Breeding season: V—VII, 1—2 broods a year. Nest: cup made of blades of grass, roots, moss and hair, usually in recess, sometimes on beam under roof. Eggs: 5, white, with grey and russet spots, incubated 12—13 days (and nestlings fed 13—14 days) by ♂ and ♀. Diet: flying insects, berries. Migrant. Migration: IV—V and VIII—IX.

2 **PIED FLYCATCHER** *(Ficedula hypoleuca)* Smaller than Sparrow. ♂ in nuptial plumage vivid black and white. Winter plumage = ♀ (grey-brown back). Juvenile: like ♀, but spotted back. Downy plumage dark grey. Voice: call-note 'whitwhit, tic, tschist', etc., song stuttering 'zee-it, zee-it'. Nests in old, open woods and parks. Breeding season: V—VI, 1 brood a year. Nest: in tree hole or nest-box, made of blades of grass, rootlets, leaves, hair. Eggs: 5—8, plain bluish green, incubated 14—16 days by ♀. Nestlings fed 14—17 days by ♂ and ♀. Diet: flying insects, in late summer also berries. Migrant. Migration: IV—V and VIII—IX.

3 **COLLARED FLYCATCHER** *(Ficedula albicollis)* Closely resembles Pied Flycatcher. ♀ in nuptial plumage has pure white nuchal band and rump. Winter plumage = ♀, greyish brown, faint neck band, but distinctly larger white area on wings. Juvenile and downy plumage: like Pied Flycatcher. Voice: different from Pied Flycatcher's, call-note 'hees' or 'zig', song series of high whistling tones, sounds like 'tsit-tseet-tsit-siu-seet'. Rare nester in open deciduous woods. Breeding season: V—VI, 1 brood a year. Nest, eggs and care of brood as for Pied Flycatcher. Diet: flying insects. Migrant. Migration: IV—V and VIII—IX.

1

2

3

1 juv
2 ♂ W
1
2 ♀
2 ♂
2 ♂
3 ♂
3 juv
3 ♂
3 ♀

Order: Passerines - Passeriformes **Family: Flycatchers -** Muscicapidae

1 **RED-BREASTED FLYCATCHER** *(Ficedula parva)* Small, superficially resembles Robin. ♂ in nuptial plumage has russet throat and breast and black and white tail. Winter and juvenile like ♀, without red throat. Behaves like other flycatchers, sits erect, twitches wings and tail, makes short flights to catch insects. Voice: call-note high, disyllabic 'hwee-a', song slow, glissando 'tink-tink-tink-eida-eida-eida-hwee-da-hwee-da', reminiscent of song of Willow Warbler. Rare nester in tall beechwoods. Breeding season: V—VI, 1 brood a year. Nest: in recess, as other flycatchers. Eggs: 5—7, whitish, covered with minute rust brown speckles, incubated about 15 days by ♀. Young fed about 14 days by ♂ and ♀. Diet: insects. Migrant. Migration: V and VIII—IX.

Order: Passerines - Passeriformes **Family: Thrushes -** Turdidae

2 **STONECHAT** *(Saxicola torquata)* Smaller than Sparrow. ♂: black head and back, white scapular spot, rust brown breast. ♀: brown back, black-striped head and throat. Juvenile: spotted back. Downy plumage brownish grey. Mostly sits erect on top of tree or bush, or on telephone wires, often seen also on ground. Voice: warning call 'wheet-tsack-tsack-tsack', song quickly reiterated scraping tones, resembles Black Redstart's. Nests on sunny hills with plenty of bushes. Breeding season: IV—VII, 2 broods a year. Nest: in depression in ground, hidden by vegetation, made of blades of grass, rootlets, moss and dead leaves. Eggs: 5—6, greenish, with small russet spots, incubated 14—15 days by ♀. Nestlings fed 12—14 days by ♂ and ♀. Diet: insects. Partial migrant. Migration: II—III and X.

3 **WHINCHAT** *(Saxicola rubetra)* Distinguished from similar Stonechat by brown head, broad white eye stripes and completely light, rust-coloured underside. Base of tail white at sides. ♀: duller colouring, no white area on wings. Juvenile: spotted back and breast. Downy plumage dark grey. Voice: call-note explosive 'tic-tic', short song consists of a few whistles and scraping tones, often includes imitations of other songbirds' motifs. Nests in meadow country with thickets and in mountain meadows. Likes to sit on highest bushes. Breeding season: V—VI, 1 brood a year. Nest: like Stonechat's. Eggs: 5—6, dark bluish green, incubated 14 days by ♀. Nestlings fed 11—15 days by ♂ and ♀. Diet: insects. Migrant. Migration: IV—V and VIII—IX.

1

2

3

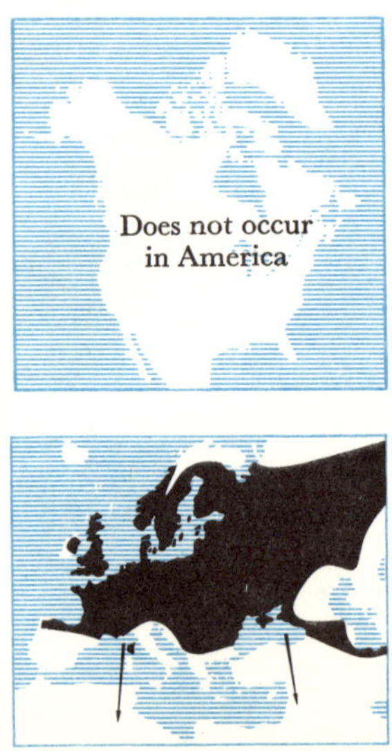

1 ♂
1 ♀
2 juv
2 ♀
2 ♀
3 ♀
3 ♂
3 juv

1 **REDSTART** *(Phoenicurus phoenicurus)* Size of Sparrow. Conspicuous for frequent 'curtsies' and quivering tail movements. ♂: white forehead, black throat and cheeks, russet breast. ♀: greyish brown. Juvenile: spotted. Downy plumage dark grey. Voice: call-note explosive 'wheet, tucc-tucc-tucc', rather short song contains a few tuneful, together with harsh, tones, always begins with one high-drawn and two deeper notes, 'he, tra-tra'. Nests in deciduous and mixed woods, gardens, parks, villages. Breeding season: V—VII, 2 broods a year. Nest: in tree hole, nest-box, hole in wall, etc., made of straw, leaves and roots, lined with hair and feathers. Eggs: 5—7, plain greenish blue, incubated 13—14 days by ♀. Nestlings fed 12—14 days by ♂ and ♀. Diet: insects, berries. Migrant. Migration: IV and IX—X.

2 **BLACK REDSTART** *(Phoenicurus ochruros)* Same form and movements as Redstart. ♂: coal black and ash grey, white area on wings, quivering russet tail. ♀: dark grey, no white area on wings, altogether darker than similar female Redstart. Juvenile: dark brown, almost unspotted. Downy plumage dark grey. Voice: song 4—5 high, stuttering notes of same pitch, a hissing sound and 2—3 more warbles, not very tuneful, but persistent, sung on roofs after dusk. Common nester in human communities, stone-quarries and rocky country, also occurs high up in mountains. Breeding season: IV—VII, 2 broods a year. Nest: in rock crevice, hole in wall, on roof-beam, sometimes in room, made of blades of grass and roots. Eggs: 5—6, white, incubated 13 days by ♀. Young fed 12—17 days by ♂ and ♀. Diet: insects, berries. Partial migrant. Migration: III and X.

3 **ROBIN** *(Erithacus rubecula)* Almost size of Sparrow, distinguished by red breast and cheeks and plain, dark brown tail. Hops about on longish, thin legs, curtsies, lets wings droop slightly. ♂ = ♀. Juvenile: light-spotted back. Downy plumage brownish black. Voice: call-note metallic 'tic-tic-tic', often repeated, song short series of clear tones with melodious trills, usually sung from high-topped tree. Common forest-dweller. Breeding season: IV—VI, 2 broods a year. Nest: in hole in ground, hollow tree stump or side of ditch, made of moss and dead leaves. Eggs 5—6, yellowish, with fine, yellowish brown speckles, incubated 14 days by ♀. Nestlings fed 12—15 days by ♂ and ♀. Diet: insects, spiders, worms, berries. Partial migrant. Migration: III and IX—X.

1

2

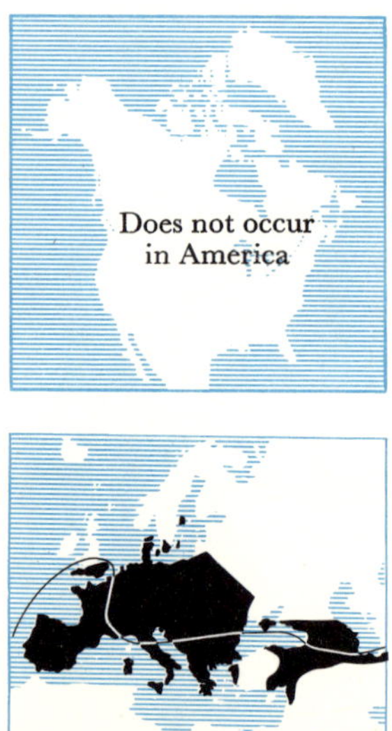

3

1 ♂
1 ♀
1 juv
2 ♀
2 ♂
3 ♂
3
3 juv

1 **NIGHTINGALE** *(Luscinia megarhynchos)* Size of large Sparrow. Plain brown, with rust brown tail. ♂ = ♀. Juvenile spotted. Downy plumage greyish black. Spends much time on and near ground, hops on long legs like typical thrush, with drooping wings and erect tail. Voice: call-note 'wheet', warning call 'kerr'. Famous song loud and melodious, consists of very varied, warbling, throbbing and swelling, fluty phrases combined in strict pattern. Sings regularly also at night. Inhabits riverside woods with thick undergrowth, parks, gardens and dry deciduous woods. Breeding season: V—VI, 1 brood a year. Nest: on ground, made of dry leaves, blades of grass, roots, moss. Eggs: 4—6, olive brown to coffee-coloured, incubated 13 days by ♀. Young fed 11—12 days by ♂ and ♀. Diet: insects, berries. Migrant. Migration: IV—V and VIII—IX.

2 **THRUSH NIGHTINGALE** *(Luscinia luscinia)* Like Nightingale, distinguished by brown-marked breast and generally darker colouring. Voice: resembles Nightingale's, but song more like Song Thrush's, lacks swelling and slurred phrases. Inhabits similar country, but prefers damper places, e.g. marshy woods, swamps with willow thickets, sunken ground with pools, river valleys, etc. Distribution also virtually complementary to Nightingale's. Care of brood and diet: largely as for Nightingale. Migrant. Migration: IV—V and VIII.

3 **BLUETHROAT** *(Luscinia svecica)* Similar to Robin in size, form and movements, but ♂ in nuptial plumage has bright blue throat and ♀ irregular black markings on breast. Base of tail rust brown. In central and eastern Europe, ♂♂ have white spot in centre of blue throat, in Scandinavia russet spot (White- and Red-spotted Bluethroat). Juvenile: lines down body. Downy plumage slate grey. Voice: warning call short 'tac', song introduced with 'deep-deep-deep', followed by series of warbling, churring and whistling tones. Nests in marshes with extensive willow thickets. Breeding season: IV—VI, 1 brood a year. Nest: on ground, made of leaves, blades of grass, roots. Eggs: 5—6, thickly covered with rusty yellow speckles, incubated 13 days. Young fed 13—14 days. Diet: insects, berries. Migrant. Migration: III—IV and VIII—IX.

1

2

3

1
1 juv
3 ♂ with white spot
2
3 ♂
3 ♀
3 ♂ with red spot
3 juv

1 **WHEATEAR** *(Oenanthe oenanthe)* Size of Sparrow. ♂ in nuptial plumage grey, ear-coverts, wings and end of tail black. ♀ (and ♂ in winter plumage) less vivid, underside ochre yellow. Juvenile: spotted. In all plumages rump and base of tail white (especially conspicuous in flight). Downy plumage dark grey. Spends much time on ground, likes to sit on large boulders, hops and curtsies in typical thrush manner. Voice: call-note repeated short 'weet' or 'weet-chack', song a short, forced twitter. Nests in rocky country, quarries, pastures with boulders, debris, rocky mountains above forest belt. Breeding season: V—VI, 1 brood a year. Nest loosely made of blades of grass and roots. Eggs: 5—6, light blue, usually unspotted, incubated 14 days. Nestlings fed about 15 days. Diet: insects, spiders, small snails. Migrant. Migration: III—IV and IX—X.

2 **BLACK-EARED WHEATEAR** *(Oenanthe hispanica)* Resembles Wheatear in size and markings, but has light ochre back and underside. Two colour variants: black-throated and white-throated (with black ear-coverts and lores). ♀ and juvenile: similar to Wheatear. Nests in dry, rocky country and vineyards. Breeding season: V—VI, probably 2 broods a year. Care of brood and diet: as for Wheatear. Migrant. Migration: III—IV and IX—X.

3 **ROCK THRUSH** *(Monticola saxatilis)* Size of Starling. ♂ in nuptial plumage variegated, shows white end of back and russet tail in flight. In winter plumage, bright colours concealed by light edges of feathers. ♀ and juvenile: scaly appearance, underside more yellow, back greyish brown. Downy plumage bluish grey, long down feathers. Voice: call-note sharp 'chack, chack'. Song composed of alternating, fluty, tuneful phrases, often sung during short display flight. Rare and shy nester in bare, rocky country, on sunny rock slopes, sometimes in vineyards. Breeding season: V—VI, 1 brood a year. Nest: loose litter of stems, roots, moss in crevice or between boulders. Eggs: 4—5, plain greenish blue, incubated 13—15 days by ♀. Nestlings fed 13—16 days by ♂ and ♀. Diet: insects, spiders, snails, berries. Migrant. Migration: IV—V and VIII—IX.

1

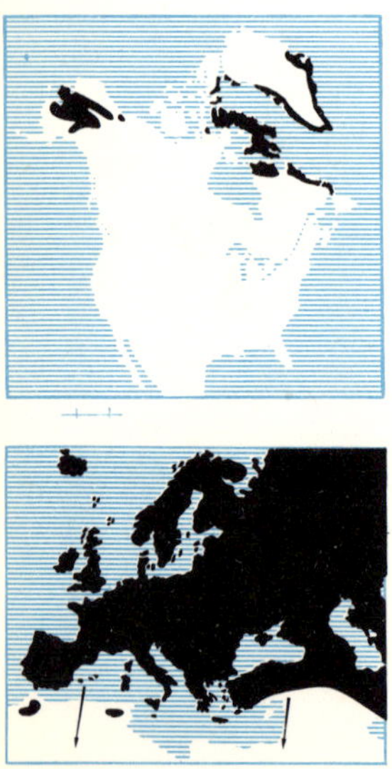

2

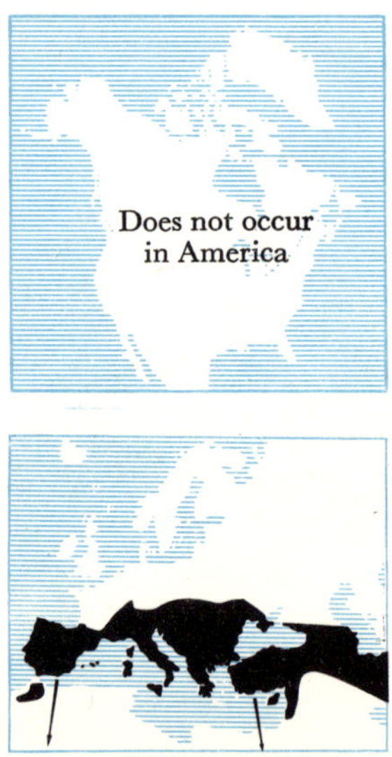

3

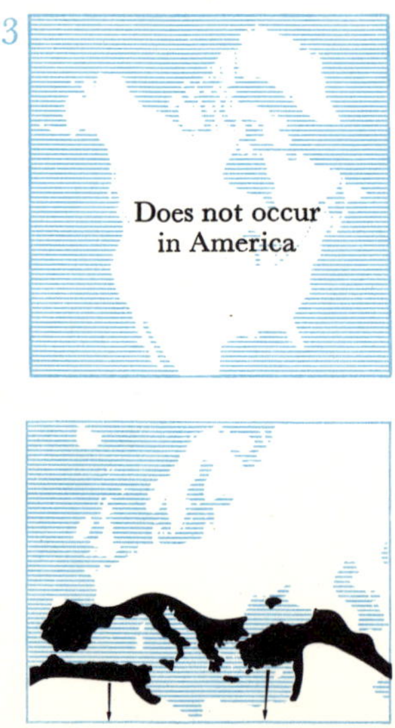

1 juv
1 ♂ S
1 ♀
2 ♀
2 ♂ S
black-throated form
2 ♂ S
2 ♂
3 ♂
3 ♂
3 ♀

1 **BLUE ROCK THRUSH** *(Monticola solitarius)* Thrush size. ♂: slate to cobalt blue, beak and feet black. ♀ and juvenile: brownish grey. Brownish throat, scale-marked underside. Voice: similar to Rock Thrush's, but shorter and less varied. Rare nester in Mediterranean region, in dry, rocky country (like Rock Thrush). Breeding season: IV—V, 1 brood a year. Nest: like Rock Thrush's. Eggs: 5, plain light blue. Little known of nest biology. Diet: insects, spiders, lizards, berries. Partial migrant. Migration: IV and IX.

2 **MISTLE THRUSH** *(Turdus viscivorus)* Slightly larger than Blackbird. Might be mistaken for large Song Thrush, but back greyer, large droplet spots extend to belly and undersurface of wings whitish, i.e. not yellow (Song Thrush), or russet (Redwing). ♂ = ♀. Juvenile: light-spotted back. Downy plumage brownish white. Voice: fluty song resembles Blackbird's, but with much simpler motifs. Warning call highly characteristic rattling sound. Common nester in large forests, in western Europe also often in village gardens and parks (like Blackbird). Breeding season: IV—VI, 2 broods a year. Nest: made of stalks, brushwood, roofs, usually high up in tree. Eggs: 4—5, light blue with russet spots, incubated 14 days by ♀. Nestlings fed 14—16 days by ♂ and ♀. Diet: insects, snails and slugs, worms, berries. Partial migrant. Migration: II—III and X—XI.

3 **SONG THRUSH** *(Turdus philomelos)* Smaller than Blackbird. Brown back, creamy, thickly spotted breast, undersurface of wings yellow. ♂ = ♀. Juvenile: spotted back. Downy plumage yellowish brown. Voice: call-note 'sip'. Song loud and full, with various motifs, each repeated about three times. Common forest bird, also occurs in parks and gardens. Breeding season: IV—VII, 2 broods a year. Nest: strong structure made of dry stalks, blades of grass and moss, walls of deep cup smoothed with fine wood pulp. Usually positioned low down, in dense shrub or conifer. Eggs: 4—6, greenish blue with a few black speckles, incubated 12—13 days by ♀. Nestlings fed about 14 days by ♂ and ♀. Diet: as for Mistle Thrush. Partial migrant. Migration: III and IX—X.

1

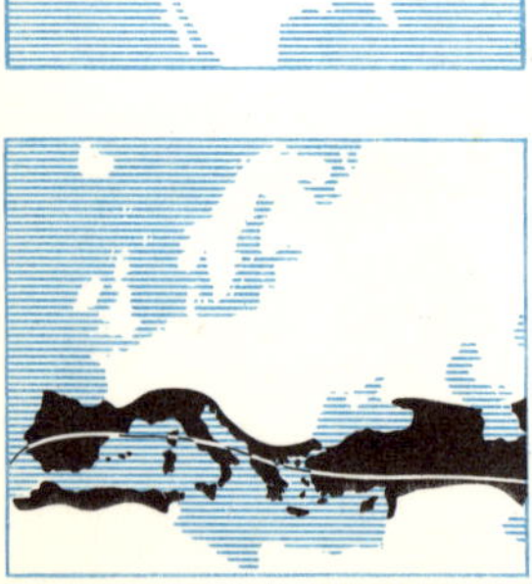

2

3

1 ♀
1 ♂
2 juv
2 ♂
2 ♂
3

1 **FIELDFARE** *(Turdus pilaris)* Size of Blackbird. Head and rump ashen, back brown, tail black. Spotted russet breast. ♂ = ♀. Juvenile: brown head and rump. Downy plumage light brown. Voice: call-note, frequently heard, loud 'tchak-tchak-tchak'; song forced twitter, often sung during flight. Nests in patchy woods and in copses, beside ponds with old trees, in birch stands, etc., often in small colonies. Breeding season: IV—VI, 1—2 broods a year. Nest: resembles Blackbird's; usually in deciduous tree or pine. Eggs and care of brood: as for Blackbird. Diet: insects, worms, berries. Partial migrant and winter bird. Migration: III—V and X—XI.

2 **REDWING** *(Turdus iliacus)* Resembles slightly larger Song Thrush, except for light eye stripes and russet axillaries and sides. Recognizable in flight by russet undersurface of wings. ♂ = ♀. Juvenile: spotted back, without russet sides. Downy plumage brownish grey. Voice: call-note unmistakable, protracted 'see-ip'; migrating flocks chatter loudly like starlings as a rule. Song low and chattering, with a few Song Thrush motifs. Nests in light woods, often in marshy country; in winter occurs in open country with scattered woods and groups of trees. Breeding season: V—VII, 2 broods a year. Nest and eggs: similar to Blackbird's. Eggs incubated 14—15 days and nestlings fed 11—14 days by ♂ and ♀. Diet: as for Song Thrush. Partial migrant. Migration: III—IV and X—XI.

3 **RING OUZEL** *(Turdus torquatus)* Size of Blackbird. ♂: black, with white-edged feathers; light area on wings, white, crescent-shaped band on crop. ♀: dark brown, light crop band. Juvenile: brown, vividly spotted underside. Downy plumage brown. Voice: warning call blackbird-like 'tac-tac-tac'; song contains repeated motifs, mostly 'dru-dru-dru', but less tuneful than Song Thrush's, with croaking tones. Nests in mountain forests and subalpine stunted timber zone, spends much time on ground, scours meadows for food. Darker northern variety occurs in western Europe as regular migrant. Breeding season: IV—VI, 1—2 broods a year. Nest: like Blackbird's. Eggs: 4—5, bluish green, densely covered with rust brown spots, incubated 12—14 days (and young fed 14—16 days) by ♂ and ♀. Diet: as for Blackbird. Partial migrant. Migration: III—IV and IX—XI.

1

2

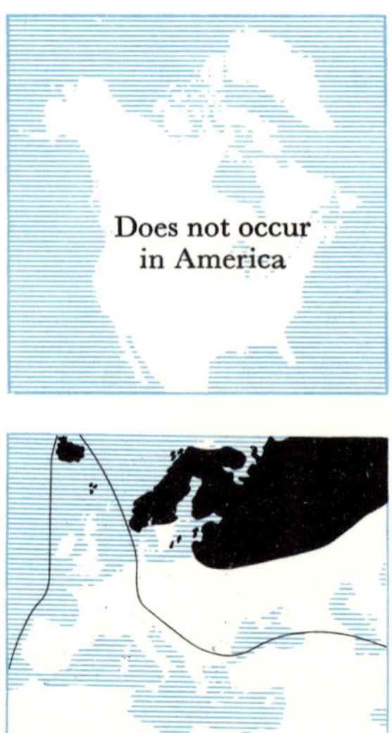

3

1 juv
1 ♂
2
2
2 juv
3
3 ♀
3 ♂
north-European form
3 ♂
central-European form

Order: Passerines - Passeriformes **Family: Thrushes** - Turdidae

1 **BLACKBIRD** *(Turdus merula)* ♂: black, with orange yellow beak and orbital ring. ♀: brown, with whitish throat and spotted breast. Juvenile: light streaks down back. Downy plumage brownish grey. Voice: call-note 'tchook, tchook, tchook' and 'tsee', warning call noisy 'chick-chick-chick-chick'. Song full-throated and fluty. Common nester in woods, parks, gardens and human communities, even cities. Breeding season: IV—VII, 2—3 broods a year. Nest: quite large, made of stalks, blades of grass and rootlets bound with mud, in bush or tree or on building. Eggs: 4—6, bluish green, thickly covered with russet speckles, incubated 13—14 days by ♀. Young fed by ♂ and ♀, 13—15 days in nest and 2 more weeks after fledging. Diet: insects, earthworms, snails, slugs, berries, fruit. Partial migrant. Migration: III—IV and X—XI.

Order: Passerines - Passeriformes **Family: Bearded Tits** - Paradoxornithidae

2 **BEARDED TITMOUSE** *(Panurus biarmicus)* Smaller than Sparrow, long tail. ♂: ashen head, vivid moustachial stripe. ♀: brown head. Juvenile: back and tail largely black. Downy plumage not developed, pearly, photo-reflective papillae in beak cavity. Voice: call-note a twanging 'tching tching', when excited a churring 'tchirr-irr irr'. Song: soft twitter. Climbs nimbly among reeds, flies low and jerkily over reed-beds. Rare and sporadic nester in reed-beds in a few pond and lake regions. Breeding season: IV—VII, 2 broods a year. Nest: deep cup made of reed blades on bent reeds over water. Eggs: 5—7, white, with black spots and streaks, incubated 12 days (nestlings fed 10—13 days) by ♂ and ♀. Diet: insects, in winter also seeds. Resident and nomadic bird.

Order: Passerines - Passeriformes **Family: Long-tailed Tits** - Aegithalidae

3 **LONG-TAILED TIT** *(Aegithalos caudatus)* Small, with round body and extremely long, gradated tail. Head white (northern Europe) or with wide, curved supraorbital stripe (western Europe), mixed populations in central Europe. ♂ = ♀. Juvenile: blackish brown cheeks and nape, lacks reddish shoulders. Downy plumage mousy. Voice: thin 'tsee-tsee' calls, rolling 'tsirrr' or 'tsirrrup'. Mostly consorts in parties, scours tips of twigs for food, nimbly climbing or hanging upside down. Common nester in parks, gardens and open deciduous and mixed woods. Breeding season: IV—VI, 2 broods a year. Nest: thick-walled, closed ovoid with side entrance, made of moss, plant fluff and lichen. Eggs: usually 7—10, white, faintly spotted, incubated 12 days. Nestlings fed 15 days by ♂ and ♀. Diet: small insects (aphids). Partial migrant. Migration: II—IV and IX—XI.

1

2

3

1 juv
1 ♂
1 ♀
2 ♂
2 ♀
2 juv
3 stripe-headed form
3 juv
3 white-headed form

Order: Passerines - Passeriformes **Family: Penduline Tits** - Remizidae

1 **PENDULINE TIT** *(Remiz pendulinus)* Much smaller than Sparrow. Greyish white head with black mask, russet back. ♂ = ♀. Winter plumage and juvenile: no mask, brown head. Voice: call-note very thin, protracted, unmistakable 'seee'. Song: soft twitter. Sporadic nester in well-watered country, riverside woods, moors, ponded regions. Climbs nimbly among reeds and in trees. Breeding season: IV—VI, 1—2 broods a year. Nest: closed pouch with side entrance, made of soft fluff or wool thickly interwoven with fibres, usually suspended from tips of twigs low down over water. Eggs: 6—8, white, incubated 14 days (and nestlings fed 15—20 days) by ♀. Male largely builds nest. Diet: insects, in winter seeds. Nomadic bird.

Order: Passerines - Passeriformes **Family: Titmice** - Paridae

2 **CRESTED TIT** *(Parus cristatus)* Small, with pointed crest, black-framed face and brownish grey back. ♂ = ♀. Juvenile: faint facial markings, short crest. Downy plumage dark grey. Voice: call-note rolling 'tzee-tzee-choor-r-r'. Song inconspicuous, intermingled with characteristic call-notes. Nests in conifer woods, in southern Europe also in oak stands. Breeding season: IV—VI, 2 broods a year. Nest: in tree hole or nesting-box, made of moss, lichen and blades of grass, lined with hair. Eggs: 7—10, white, with vivid red spots at blunt end, incubated 15—18 days by ♀. Young fed by ♂ and ♀, 20 days in nest and about 14 days after fledging. Diet: insects, insects' eggs, in winter also seeds. Resident and nomadic bird.

3 **COAL TIT** *(Parus ater)* Small, with black head, sharply defined white cheeks and large white nuchal spot. Underside whitish, sides pale brown. ♂ = ♀. Juvenile: duller colouring, dark brown instead of black. Downy plumage smoky grey. Voice: soft, thin 'tsee-tsee', sometimes strung together to 'tsitsitsitsitsit'. Song: high-pitched 'seetoo-seetoo-seetoo', etc. Common nester in conifer woods (in mountains up to tree limit). Breeding season: IV—VI, mostly 2 broods a year. Nest: cup made of moss, roots, blades of grass, lichen and wool, in tree hollow, nest-box or hole in ground. Eggs: 8—10, white with reddish speckles, incubated 14—16 days by ♀. Young fed by ♂ and ♀, 16—17 days in nest and some time after leaving nest. Diet: insects, in winter conifer seeds. Partial migrant. Migration: IX—X.

1

2
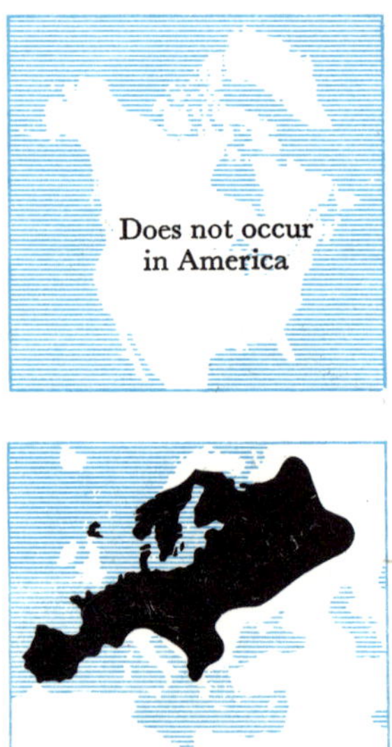

3

1 juv
1
2
3
3 juv

1 **MARSH TIT** *(Parus palustris)* Smaller than Sparrow. Greyish brown, with glossy black cap, white cheeks and small black chin spot. No light area on wings (cf. Willow Tit). ♂ = ♀ = juvenile. Downy plumage brownish grey. Voice: call-note 'pitchew'. Song mostly prattling 'schip-schip-schip' or whistling 'schüppi-schüppi-schüp', etc. Widespread nester in light deciduous woods, bush-strewn marshes, gardens and parks. Breeding season: IV—V, 1 brood a year. Nest: basin made of lichen, blades of grass and hair, usually in hole near foot of tree. Eggs: 7—10, white with a few red spots, incubated 13 days by ♀. Nestlings fed 16—18 days by ♂ and ♀. Diet: insects, insects' eggs, in winter seeds. Resident bird.

2 **WILLOW TIT** *(Parus montanus)* Closely resembles Marsh Tit, but has dull black cap (like young Marsh Tit), larger chin spot and often distinct light area on wings (light-edged primaries). Best differentiated by voice: call-note conspicuous, noisy, broad 'chay-chay' or 'chichitchay-tchay-tchay'. Song 5—6 whistling 'piū-piū-piū-piū-piū' notes of same pitch. Sporadically common bird in young mixed woods, alder swamps, marshes with thickets, in mountains up to forest limit. Breeding season: IV—VI, 1—2 broods a year. Nest: like Marsh Tit's, often in rotting tree stump. Eggs: 7—8, white with red spots, incubated 13—14 days by ♀. Nestlings fed 17—19 days by ♂ and ♀. Diet: as for Marsh Tit. Resident bird.

3 **BLUE TIT** *(Parus caeruleus)* Smaller than Sparrow. Bright blue crown and wings, light yellow underside. ♂ = ♀. Juvenile: paler. Downy plumage greyish white. Voice: mostly a shrill 'tserret-et-et-et'. Song a clear 'tsee-tsee-tseerrrr'. Common bird in parkland, gardens and open mixed woods. Nimble climber. Breeding season: IV—VI, 2 broods a year. Nest and eggs: like Great Tit's. Eggs incubated 13—15 days by ♀, nestlings fed 17—18 days by ♂ and ♀. Diet: insects, in autumn also berries, in winter mainly oily seeds. Partial migrant. Migration: III—IV and IX—XI.

1

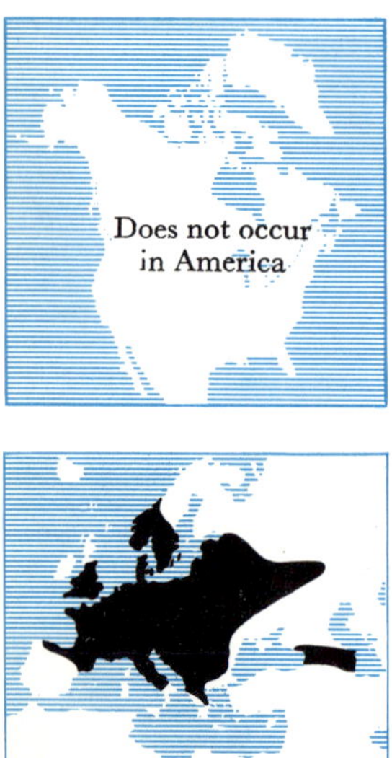

2

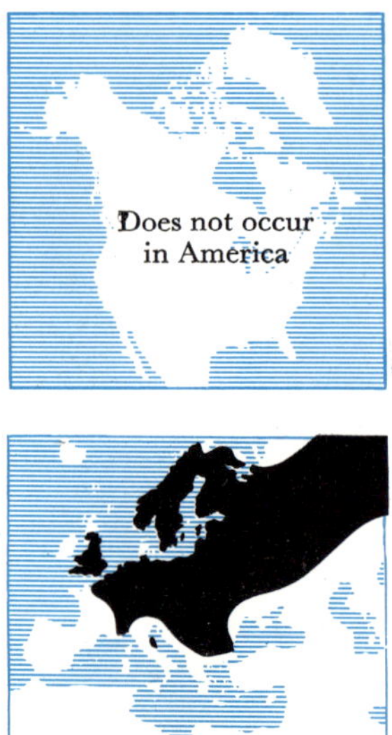

3

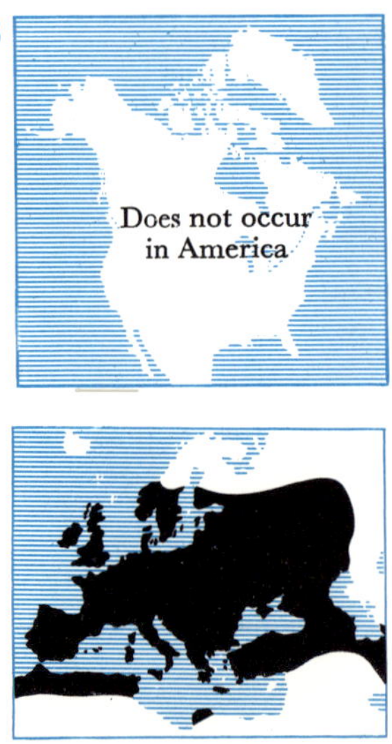

1
1
2
3
3 juv

Order: Passerines - Passeriformes **Family: Titmice** - Paridae

1 **GREAT TIT** *(Parus major)* About size of Sparrow. Brightly coloured, very lively bird. Glossy black head with pure white cheeks. Yellow underside, broad black stripe down midline. ♀: thinner belly stripe. Juvenile: duller colours, practically no belly stripe. Downy plumage smoky grey. Flies jerkily, climbs nimbly in branches, often looks for food on ground. Voice: call-note clear 'tsink' and interrogative 'tsee-tūi'. Song rhythmical, sharp, metallic 'teechūwee-teechūwee', etc. Common nester in woods, gardens, avenues and parks, in mountains up to tree limit. Breeding season: IV—VI, 2 broods a year. Nest: made of moss, lichen, straw, roots, in tree hole, nest-box or some kind of recess. Eggs: 8—10, white with red spots, incubated 13—14 days by ♀. Nestlings fed 15—20 days by ♂ and ♀. Diet: insects, oily seeds, fruit. Partial migrant. Migration: III—IV and IX—X.

Order: Passerines - Passeriformes **Family: Nuthatches** - Sittidae

2 **NUTHATCH** *(Sitta europaea)* Size of Sparrow. Large head, straight, awl-like beak, short tail. Greyish blue back, ochre yellow underside. ♂ = ♀ = juvenile. Downy plumage dark grey. Voice: call-note short, resolute 'tsit', when excited 'chwit-chwit'. Song: loud, clear, piping 'twee-twee-twee', often condensed to rapid 'chūchūchūchūchū'. Mostly climbs jerkily up tree trunks, also descends head first. Common nester in woods, parks, gardens and avenues with old deciduous trees. Breeding season: IV—VI, 1 brood a year. Nest: loose litter of bark, dead leaves and fibres, in tree hole, nesting box or hole in wall; entrance regularly made smaller with clayey mud to fit bird's girth. Eggs: 6—8, white with russet spots, incubated 13—17 days by ♀. Nestlings fed about 24 days by ♂ and ♀. Diet: insects, oily seeds, sometimes berries. Resident bird.

3 **WALL CREEPER** *(Tichodroma muraria)* Size of large Sparrow. Long, thin, curved beak, short tail. Upper surface of wings crimson, with white droplet spots on primaries. Throat and crop black. Juvenile and winter plumage: throat and crop white. Downy plumage grey. Voice: thin, piping 'diu-dia' call notes, etc. Song whistling 'zizizizui'. Frequents bare, precipitate rocks, rarely masonry. Climbs with half-spread wings, fluttering. Rare nester at high altitudes in large mountain ranges, in winter occasionally also appears in the lowlands. Breeding season: V—VI, 1 brood a year. Nest: made of moss, lichen, rootlets, hair, in crevice. Eggs: 4—5, white with a few russet spots, incubated 18—19 days. Hatching period 21—23 days. Diet: insects, spiders. Partial migrant. Migration: III and IX—X.

1

2

3

1 juv
1 ♂
2 S
3 W

Order: Passerines - Passeriformes **Family: Tree-creepers** - Certhiidae

1 **TREE CREEPER - BROWN CREEPER** *(Certhia familiaris)* Small and dainty. Thin, curved beak, long, supporting tail. Tobacco brown back, gleaming white underside (cf. Short-toed Tree Creeper). ♂ = ♀ = juvenile. Downy plumage blackish grey. Voice: soft 'tsee, tsee'. Song louder and longer than Short-toed Tree Creeper's, something like 'tsee-tsee-tsizzi-tsu-izzirr-ee', with slight resemblance to Wren's song (rolling trill). Climbs tree trunks in jerks, flies down again to foot of next tree. Common nester in fir-woods and mixed woods. Breeding season: IV—VII, 1—2 broods a year. Nest: made of small twigs, leaves and blades of grass, in crevice in tree, woodpile, under roof, etc. Eggs: 5—7, whitish with red speckles, incubated 14—15 days by ♀. Young fed by ♂ and ♀. Diet: insects, insects' eggs, supplemented by seeds. Resident bird.

2 **SHORT-TOED TREE CREEPER** *(Certhia brachydactyla)* Sometimes distinguished from very similar Tree Creeper by dull grey belly, browner sides, more mousy back, less distinct eye stripe and longer, more curved beak. Best differentiated by voice: call-note energetically repeated 'teet'. Song short, thin, piping 'teet, teet, teerrooititt', with no trill. Common bird in mixed woods, avenues, gardens and parks with old deciduous trees. Breeding season: V—VII, 1 brood a year. Nest: made of twigs, blades of grass and leaves, with softly lined cup, in crack in tree, woodpile or under roof. Eggs, care of brood and diet: as for Tree Creeper (spots on eggs usually larger). Resident bird.

Order: Passerines - Passeriformes **Family: Buntings** - Emberizidae

3 **SNOW BUNTING** *(Plectrophenax nivalis)* Size of large Sparrow. Light plumage, large, snow-white areas on wings. ♂ in nuptial plumage white and black, with white head. ♀: white and brown. Winter plumage: white and light brown, ♂ = ♀. Distinctive pure white, black-tipped wings in flight (similar to Snow Finch). Juvenile: rust brown wings. Downy plumage dark grey. Voice: call-note trilling 'trree'. Short, melodious song, 'turi-turi-turi-tetitui', often heard during display flight. Nests in rocky tundras. Common winter bird, often in large flocks near coast, less frequently inland in fields. Ground-dweller, tripping gait. Breeding season: VI—VII, 1 brood a year. Nest: made of blades of grass, moss and lichen, on ground. Eggs: 5—6, whitish with dark markings, incubated 12—14 days by ♀. Diet: insects, seeds. Migrant. Migration: III—IV and IX—X.

1

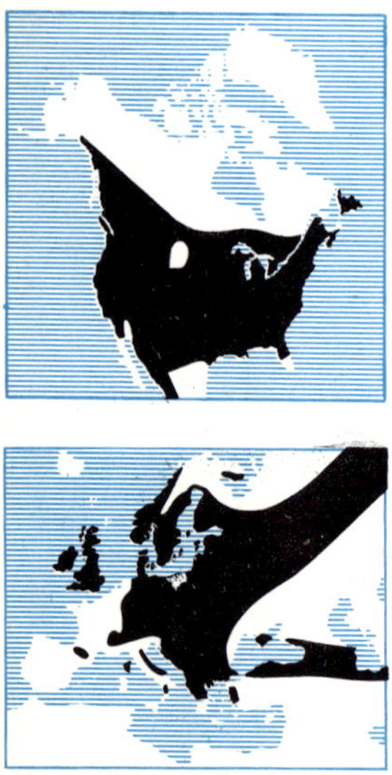

2

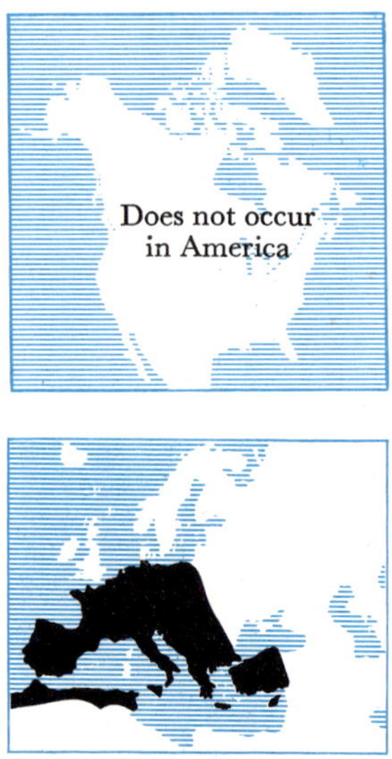

3

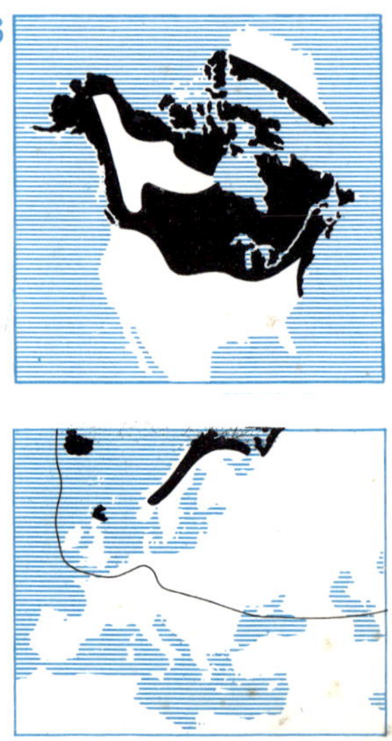

1
2
3 ♀
3 ♂
3 ♂ W

1 **CORN BUNTING** *(Emberiza calandra)* Larger than Sparrow, with short, strong beak. Greyish brown, dark-streaked. ♂ = ♀. Juvenile: browner, rusty yellow head, more thickly spotted. Downy plumage yellowish brown. Voice: call-note monosyllabic 'chip' or serial 'tseektseektseek'. Song rapid, rattling jingle, 'tseek-tseek-tseek-shnirlrlrl', sung from eminence (often telephone wires). Nests in cultivated steppes, especially cornfields and meadows with avenues and hedges. In winter roams about in flocks. Breeding season: V—VI, 1—2 broods a year. Nest: made of blades of grass and roots, lined with hair, on ground (in grass). Eggs: 4—5, pinkish yellow or whitish, with dark brown spots and scribbles, incubated 12—13 days by ♀. Hatching period 9—12 days, young leave nest before fledged, fed by ♂ and ♀. Diet: seeds, insects. Partial migrant. Migration: III and X.

2 **YELLOWHAMMER** *(Emberiza citrinella)* Size of large Sparrow. Golden yellow head and underside, rust brown rump, russet stripes on breast and sides. ♀ and juvenile: duller, more striped. Downy plumage smoky grey. Voice: call note 'chip' and 'twitzurrr'. Song series of 5—6 notes of same pitch, in regular time, with protracted and usually higher finish note, 'tiutiu-tiutiutiuk-swee'. Hops about on ground, flicks tail. Common nester in copses, at margin of woods, in tree-nurseries. In winter appears in flocks near human communities. Breeding season: IV—VII, 2 broods a year. Nest: made of blades of grass, roots and leaf fragments, on or near ground, in grass or bushes. Eggs: 4—5, whitish, with brown spots and scribbles, incubated 12—14 days (and young fed same duration) by ♂ and ♀. Diet: seeds, berries, insects. Partial migrant. Migration: III—IV and IX—XI.

3 **CIRL BUNTING** *(Emberiza cirlus)* Resembles Yellowhammer in form and size. ♂: black crown and throat, grey breast-band. ♀ and juvenile: distinguished from Yellowhammer by darker head and olive brown rump. Downy plumage greyish brown. Voice: call-note thin 'zit' or 'sip'. Song: jingling 'zirlrl', rather like song of Lesser Whitethroat. Nests in same type of country as Yellowhammer, but sunnier. In winter appears in parties on farms. Breeding season: IV—VII, 2 broods a year. Nest and eggs: like Yellowhammer's. Eggs incubated 11—12 days by ♀. Nestlings fed 11—13 days, almost entirely by ♀. Diet: as for Yellowhammer. Resident bird.

1

2

3

1
2 ♂
2
2 ♀
2 juv
3 ♂
3 ♀

1 **ORTOLAN BUNTING** *(Emberiza hortulana)* Size of Sparrow. Pink beak. ♂: grey head, yellow orbital ring, yellow throat; unspotted, cinnamon underside. ♀ and juvenile: lighter, with streaked crop, dark moustachial stripe, light orbital ring. Downy plumage light grey. Voice: call-note short 'teu' and 'tsip'. Song similar to Yellowhammer's, but soft and melancholy, 'zeu-zeu-zeu-trrull'. Sings from heights. Locally common bird in warm, fertile plains and hills with cornfields, fruit-tree avenues, tree-lined paths and vineyards. Breeding season: V—VII, 1—2 broods a year. Nest: made of blades of grass, roots and dry leaf fragments, softly lined with hair, usually in depression in ground. Eggs: 4—5, whitish, with a few brownish black spots, incubated 11—13 days by ♀. Nestlings fed 10—13 days by ♂ and ♀. Diet: seeds, insects. Migrant. Migration: V and VIII.

2 **ROCK BUNTING** *(Emberiza cia)* Size of Sparrow. Light grey, black-striped head, cinnamon underside and rump. ♀: duller, with lightly spotted sides. Downy plumage grey. Voice: call-note sharp 'tzit' or protracted 'seep'. Song: 'zi-zi-zi-zirrr', rather like Serin's or Reed Bunting's. Nests on warm, rocky slopes with bushes and in vineyards, also in mountains. Breeding season: V—VII, 1—2 broods a year. Nest: like Yellowhammer's in structure, built on ground, under clump of grass, between stones, in wall. Eggs: 4—5, greyish white, with black dots and lines, incubated 12—13 days by ♀. Nestlings fed 10—13 days by ♂ and ♀. Diet: seeds, insects. Resident and nomadic bird.

3 **REED BUNTING** *(Emberiza schoeniclus)* Size of Sparrow, plumage without any yellow. ♂ in nuptial plumage: sharply defined black head and throat, white neck band and white moustachial stripe, white outer tail feathers. ♂ in winter plumage, ♀ and juvenile: greyish white throat, black and white moustachial stripes, white eye stripe, striped underside. Voice: call-note sharp 'tseek'. Song: a staccato, stuttering 'tseek-teet-tai-tississisk'. Nests on marshy ground beside ponds and rivers, on moors, in wet meadows; is confined to sedge and reed stands. Breeding season: V—VII, 1—2 broods a year. Nest: made of blades of grass, moss and leaf fragments, lined with hair, on ground, in clump of grass or willows. Eggs: 5—6, brown, with black spots and lines, incubated 12—14 days, mostly by ♀. Nestlings fed 11—13 days by ♂ and ♀. Diet: insects, molluscs, crustaceans, seeds. Partial migrant. Migration: III and X.

1

2

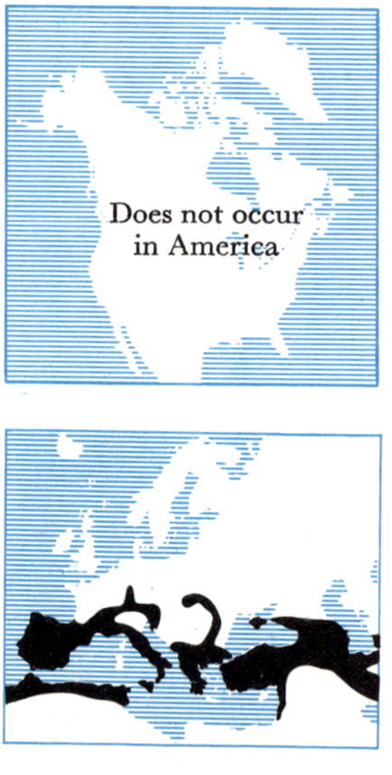

3

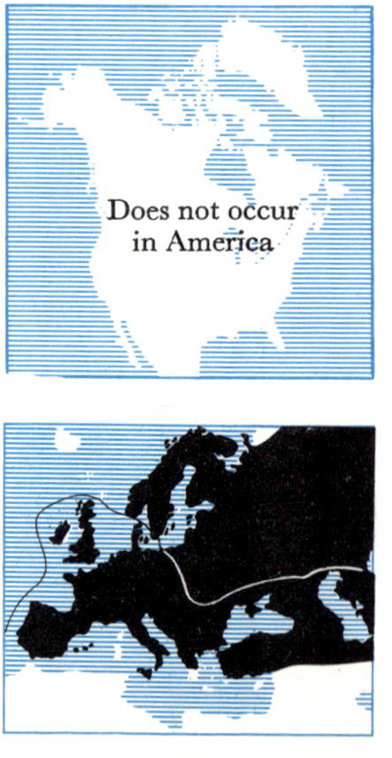

1 ♀
1 ♂
2 ♀
2 ♂
3 ♀
3 ♂
3 ♂ W

1 **BLACK-HEADED BUNTING** *(Emberiza melanocephala)* Size of large Sparrow. ♂: black cap (in autumn brown), yellow underside, brown, unstriped back. ♀ and juvenile: grey, with light, unstriped underside, yellow under tail-coverts. Voice: call-note sharp 'zitt'. Song quite tuneful, with quick trill at end, sounds like 'chit-chit-chit-eeu-chiriri'. Nests in open, bushy (and mostly hilly) country, in olive groves and vineyards. Breeding season: V—VI, 1 brood a year. Nest: in dense vegetation on ground, or in thicket near ground, made of blades of grass, rootlets, dead leaves, softly lined with hair. Eggs: 4—5, whitish, with fine, dark streaks, incubated 14 days. Hatching period unknown. Care of brood almost entirely by ♀. Diet: seeds, insects. Migrant. Migration: IV—V and VIII.

2 **LAPLAND BUNTING - LAPLAND LONGSPUR** *(Calcarius lapponicus)* Size of Sparrow. ♂ in nuptial plumage: head and throat black, nape russet. In winter, black head markings partly covered by brown feathers. ♀: white throat, striped crown, often russet nape. In winter resembles Reed Bunting, but distinguished by lark-like, running gait and short tail. Does not perch on trees. Voice: call-note sparrow- or linnet-like 'ticky-tik-ticky-tick', on nesting site piping 'teeu' calls. Song similar to Skylark's. Nest in treeless tundra, in winter occurs in stubble fields, seldom far inland. Breeding season: V—VII, 1 brood. Nest: on ground, made of grass and moss. Eggs: 4—5, brown-speckled, incubated 14 days. Diet: insects, seeds. Migrant. Winter bird. IX—III.

Order: Passerines - Passeriformes **Family: Finches** - Fringillidae

3 **HAWFINCH** *(Coccothraustes coccothraustes)* Size of Starling. Large head, strong, conical beak, short tail. Black chin spot, blue beak (in winter horny brown). ♀: lighter. Juvenile: white chin, scaly marks on underside. Distinguished in flight by large head and beak, large white areas on wings and white-ended tail. Voice: call-note loud 'tzik' or 'tzitt'. Song slightly stuttering, interspersed with call-notes. Nests in light deciduous woods and orchards. Breeding season: V—VI, 1 brood a year. Nest: in tree, made of brushwood, roots, moss, lined with hair. Eggs: 5, light blue, with a few black spots, incubated 14 days (and nestlings fed same length of time) by ♂ and ♀. Diet: cherry-, sloe-, plum-stones, apple pips, hard seeds of various deciduous trees, insects. Resident and nomadic bird.

1

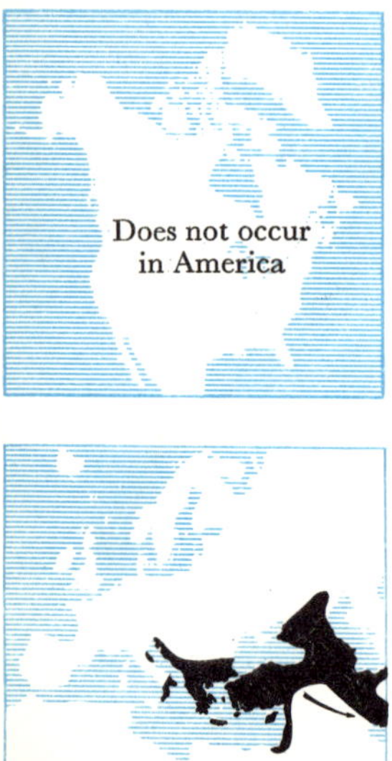

2

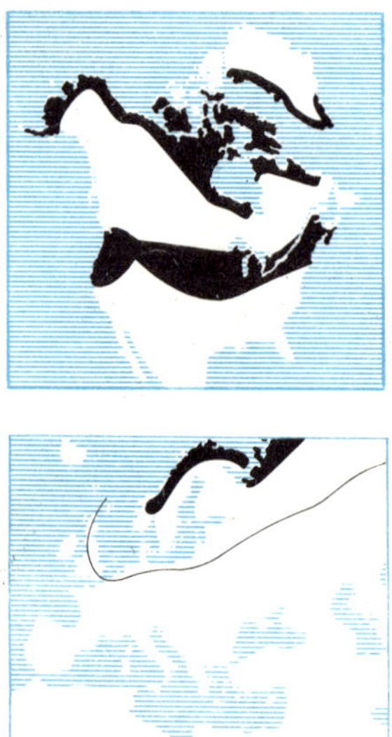

3

1 ♂ W
1 ♂
1 ♀
2 ♂
2 ♂ W
2 ♀
3
3 ♂
3 juv

1 **CHAFFINCH** *(Fringilla coelebs)* Size of Sparrow. Two distinct white wing bands and white outer tail feathers visible during flight. ♂: variegated (blue-grey head, claret underside, green rump). ♀ and juvenile: greyish brown. Downy plumage light grey. Flies in curves, trips along ground. Voice: call-notes clear 'chiuk', when flying muted 'tsip', often a vibrating 'cheeer' (rain note). Song loud, warbling, descending series of notes with characteristic di- to trisyllabic end motif. Common nester in woods and gardens, also occurs in towns. Breeding season: IV—VI, 2 broods a year. Nest: woven cup made of fibres, blades of grass, moss and lichen in tree or bush. Eggs: 4—6, mostly brown, with characteristic dark 'burn' spots, incubated 12—13 days by ♀. Nestlings fed 13—14 days by ♂ and ♀. Diet: seeds, insects. Partial migrant. Migration: III—IV and IX—X.

2 **BRAMBLING** *(Fringilla montifringilla)* Size of Sparrow. ♂ in nuptial plumage: head and front part of back deep black, rump and belly white, breast and shoulders orange. In flight shows vivid white area on anterior wing border, white band on black wings, gleaming white rump and black tail. In winter, black on head largely concealed by grey- and brown-edged feathers. ♀: head mainly brown, sides of neck grey. Juvenile: rump and belly yellowish brown. Downy plumage white. Voice: 'tsweek', when flying 'chucc-chucc-chucc'. Song unmusical and rattling, with many call-notes. Common nester in Scandinavian birch- and pine-woods, when migrating often occurs in huge flocks in fields and beech-woods. Breeding season: V—VII, 1 brood. Nest, eggs and care of brood: similar to Chaffinch. Diet: insects, seeds, berries. Migrant. Migration: III—IV and X—XII.

3 **GREENFINCH** *(Carduelis chloris)* Size of Sparrow. Olive green, with large head and strong beak. Golden yellow areas on wings and sides of tail. ♀: duller, more grey and brown. Juvenile: streaks down back and underside. Downy plumage white. Voice: call-note peculiar, screeching 'tswee' and jingling 'chichichichit'. Song, often sung during bat-like display flight, sounds like 'chichichichit-tsweerrr-teu-teu-teu-chupchupchup-djul-djul-djul . . .', etc. Common bird in copses, at edge of woods, in gardens and parks. Breeding season: IV—VI, 2 broods. Nest: cup made of blades of grass and brushwood, softly lined, in hedge, dense thicket, young conifer, etc. Eggs: 5—6, white, with dark brown spots and speckles, incubated 13—14 days by ♀. Nestlings fed 12—14 days by ♂ and ♀. Diet: seeds, berries, buds. Partial migrant. Migration: III—IV and X—XII.

1

2

3

1 ♂ W
1 ♀
1 ♂
2 ♂
2 ♂ W
2 ♀
3 ♂
3 ♀

1 **GOLDFINCH - EUROPEAN GOLDFINCH** *(Carduelis carduelis)* Smaller than Sparrow. Head markings unmistakable (red, white, black). Shows broad, golden yellow wing band and white rump in flight. ♂ = ♀. Juvenile: head plain brown, back and underside spotted, yellow wing band present. Downy plumage grey. Voice: call-note 'tswitt-witt', especially when flying. Hurried song always intermingled with call-notes. Common nester in open country with trees, village gardens, orchards. In winter occurs in flocks on barren land and roadsides with thistles and burrs. Breeding season: V—VII, 2 broods. Nest: thick-walled cup made of fine stems, blades of grass, roots, moss and lichen, high up in tree at end of branch. Eggs: 5—6, whitish with red speckles. Incubated 12—13 days by ♀. Young fed about 14 days by ♂ and ♀. Diet: seeds (especially of thistles), insects. Partial migrant. Migration: III—IV and X—XI.

2 **SISKIN** *(Carduelis spinus)* Small, yellowish green, with two yellow wing bands. ♂: black crown and chin spot. ♀ and juvenile: greyish green, no black head markings. (For possibilities of mistaken identity, see Redpoll, Citril Finch, Serin.) Voice: commonest call-note thin, protracted 'tsy-zi'. Song rapid twitter, with characteristic screech, 'deedeedleleddwee-dāāā', often sung during display flight. Common nester in northern conifer and mountain forests. Outside breeding season occurs in parties, largely in alder and birch stands. Breeding season: IV—VII, 2 broods. Nest: small, deep, thick-walled, hidden between tips of twigs of high conifer. Eggs: 5, white with russet spots, incubated 13 days by ♀. Young fed 2 weeks by ♂ and ♀. Diet: mainly pine, alder, birch seeds, also insects (during breeding season). Partial migrant. Migration: VIII—V.

3 **REDPOLL - COMMON REDPOLL** *(Acanthis flammea)* Small, greyish brown, with dark stripes down back, red forehead, black chin spot and two light wing bands. ♂: reddish breast. Juvenile: grey-brown forehead. Similar species (Siskin, Serin, Citril Finch) yellowich green (also cf. Twite and Linnet). Voice: when flying, metallic, hasty 'chuch-uch-uch-uch'. Song short trills with 'errrr' and 'chuch-uch' notes. Nests in thickets in Scandinavian birch and alder woods; in mountains at upper forest limit. In winter occurs in flocks on birches. Breeding season: V—VII, 1 brood. Nest: deep cup made of twigs, blades of grass, lichen, moss, lined with hair, usually not very high in thicket. Eggs: 5, bluish with russet spots, incubated 14 days by ♀. Young fed about 12 days by ♂ and ♀. Diet: as for Siskin. Partial migrant. Migration: X—IV.

1
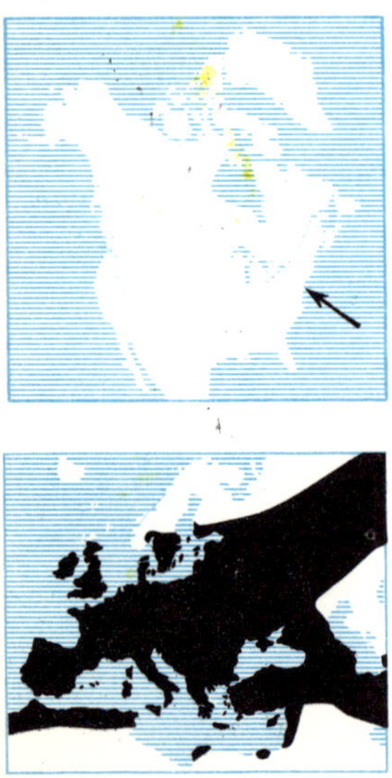

2

3
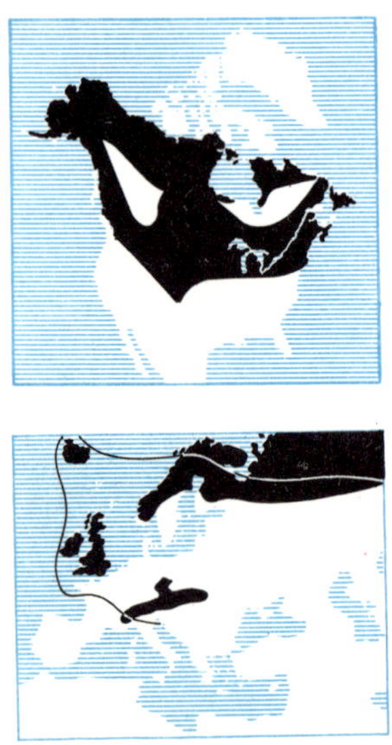

1 juv
1 ♂
2 ♀
2 ♂
3 ♂
3 ♀

1 **TWITE** *(Acanthis flavirostris)* Smaller than Sparrow. Dark brown, with cinnamon throat, dark stripes down body. Beak waxy yellow in winter. ♀: reddish rump. Juvenile: more thickly striped. Distinguished from Redpoll by absence of both chin spot and red forehead. Voice: call-note soft 'djadjadja' and loud 'chweek'. Twittering song. Nests in grassy and mossy tundra. In winter occurs in flocks in fallow and stubble fields and on waste-land. Breeding season: V—VII, usually 1 brood. Nest: cup made of blades, stems, moss, lined with hairs, close to or on ground. Eggs: 5, bluish, with russet spots and speckles, incubated about 13 days by ♀. Diet: seeds, in summer also insects. Migrant. Migration: X—III.

2 **LINNET** *(Acanthis cannabina)* Smaller than Sparrow. ♂: chestnut back, crimson breast, unstriped. ♀ and juvenile: brown, striped. Downy plumage smoky grey. Forms large flocks, especially in winter. Undulating flight. Voice: call-note (flying) hard, metallic 'chichichichit'. Song hurried and whistling, with numerous call-notes. Common nester in open country with thickets, hedges and gardens. In winter appears in flocks in fallow and stubble fields. Breeding season: IV—VI, 2 broods a year. Nest: cup made of blades of grass, rootlets, fibres, softly lined with hair and fluff, in dense thicket, young conifer, hedge, etc., usually close to ground. Eggs: 5, white with russet spots, incubated 12—14 days by ♀. Nestlings fed about 2 weeks by ♂ and ♀. Diet: seeds. Partial migrant.

3 **CITRIL FINCH** *(Serinus citrinella)* Small, size of Siskin. Greenish yellow, unstriped. Crown, nape and sides of neck grey, two greenish wing bands. Wings and tail black. ♂ = ♀. Juvenile: brownish grey, with black streaks. Voice: call-note metallic 'witt-witt'. Song jingling and twittering, often sung during display flight. Nests in open, rocky, mountainous country with isolated trees and at edge of forests. Breeding season: IV—VI, 2 broods a year. Nest: cup made of blades of grass, twigs, lichen, moss, usually high up in conifer. Eggs: 4—5, bluish green with red spots, incubated by ♀ (incubation period not known). Nestlings fed 17—18 days by ♂ and ♀. Diet: chiefly seeds, few insects. Resident and nomadic bird.

1

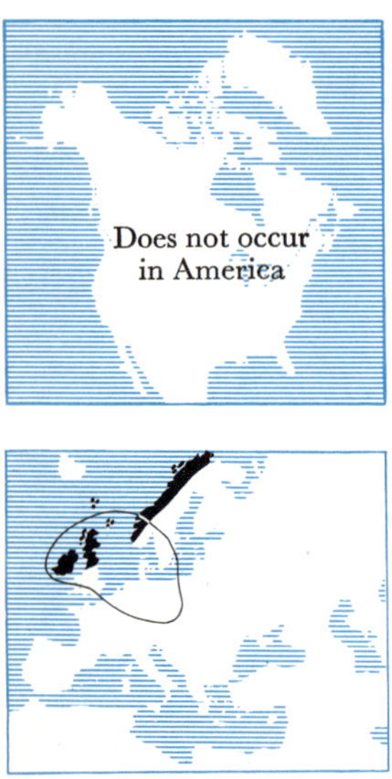

2

3

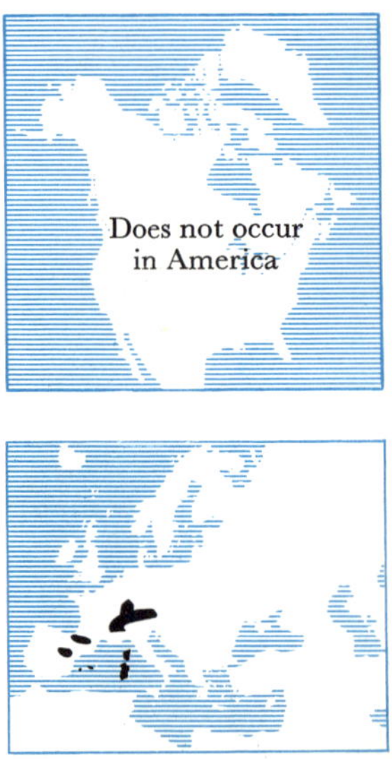

1 ♂ W
1 ♀
2 ♂
2 ♂
2 ♀
3 ♂
3 ♂
3 juv

1 **SERIN** *(Serinus serinus)* Small, greyish yellow, with short beak. ♂: bright yellow breast and rump. ♀: more striped and browner. Juvenile: brown, thickly striped, without yellow rump. Downy plumage light grey. Voice: call-note disyllabic, rattling 'tirrititt'. Song unmusical humming, sung from tip of dead branch, telephone wires or in bat-like display flight. Common nester in open country with trees, parks, gardens and at margin of woods. Breeding season: V—VII, 2 broods. Nest: thick-walled, matted cup made of stems, rootlets and lichen, lined with feathers, in small, densely foliaged deciduous tree, conifer or bush. Edge of nest always plastered with excreta. Eggs: 4, white with russet spots, incubated 13 days by ♀. Nestlings fed 14—16 days by ♂ and ♀. Diet: seeds. Partial migrant. Migration: IV and IX—X.

2 **SCARLET GROSBEAK** *(Carpodacus erythrinus)* Size of Sparrow, with thick, conical beak. ♂: head, breast and rump crimson. ♀ and juvenile: olive brown, with brown-striped breast and two light wing bands. Voice: call-note 'tiu-eek'. Song tuneful, whistling 'tiu-tiu-fi-tiu'. Rare nester in dense shrubs, mostly near water, also in mountains above forest belt. Breeding season: VI—VII, 1 brood. Nest: low down in thicket, made of thin twigs, stalks and blades of grass, smooth, rounded, well lined with hair and fluff. Eggs: 5, blue with a few blackish spots, incubated 13—14 days by ♀. Young fed 14—17 days by ♂ and ♀. Diet: seeds, buds. Summer bird: V—IX.

3 **PINE GROSBEAK** *(Pinicola enucleator)* Size of thrush, with short, thick, somewhat hooked beak. Two white wing bars. ♂: head, breast and rump rose red. ♀ and juvenile: olive green and greenish to reddish yellow. Voice: call-note fluty 'tee-tee-tew'. Song soft, fluty 'peel-peel-cheevli'. Nests in Scandinavian conifer and mixed forests. Outside breeding season consorts in small parties. Breeding season: V—VII, 1 brood. Nest: made of thin and broken conifer twigs, rootlets, blades of grass and moss, on conifer twigs, at moderate height. Eggs: 4, light blue with brownish black spots. Incubation and hatching period not known. Diet: mainly conifer seeds and buds, plus berries. Partial migrant. Accidental bird in central and western Europe.

1

2

3

1 ♀
1 ♂
2 ♂
2 ♀
3 ♀
3 ♂

1 **CROSSBILL - RED CROSSBILL** *(Loxia curvirostra)* Size of large Sparrow. Large head, crossed beak. ♂: brick red. ♀: olive brown with yellow rump. Juvenile: thickly striped underside. Downy plumage blackish grey. Often climbs about top of firs like parrot, investigating cones. Mostly consorts in parties, flies rapidly and in curves. Voice: loud, repeated 'chip-chip-chip'. Song twittering, fluty and grating tones mixed with call-notes. Widespread nester in mountain pine-woods, in irruption years also invades plains. Breeding season: whole year, chiefly I—IV. Nest: thick-walled, with deep cup, made of twigs, moss and lichen, at top of tall pine. Eggs: 4, white with russet spots, incubated 12—14 days by ♀. Nestlings fed about 14 days by ♂ and ♀. Diet: pine and fir seeds. Partial migrant with irruptive incidence.

2 **PARROT CROSSBILL** *(Loxia pytyopsittacus)* Distinguished from very similar Crossbill by stronger, deeper beak and larger head. Voice: call-note deeper than Crossbill's, sounds like 'kop-kop'. Mainly inhabits maritime pine-woods, is much more adapted for dealing with cones of maritime pines, whose seeds constitute its chief food. Care of brood: probably largely same as for Crossbill. Nest often built on maritime pine. Irregular partial migrant.

3 **TWO-BARRED CROSSBILL - WHITE-WINGED CROSSBILL** *(Loxia leucoptera)* Resembles Crossbill, but clearly distinguished, in all plumages, by two broad white bands on black wings (cf. Chaffinch). Nests in light conifer forests of northern Europe, particularly frequents larch stands. Larch seeds, together with pine seeds, form its staple diet. Voice: call-note 'chiff-chiff', softer and less emphatic than Crossbill's, song somewhat richer. Care of brood: as for Crossbill. Partial migrant. Irregular irruption bird in central and western Europe.

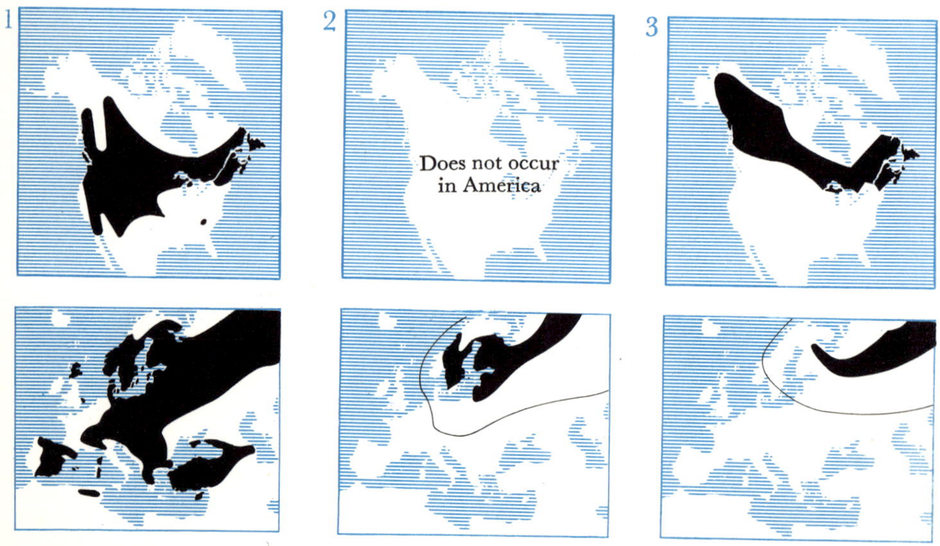

1 juv
1 ♀
1 ♂
1 ♂
2
3 ♂
3 juv

Order: Passerines - Passeriformes **Family: Finches** - Fringillidae

1 **BULLFINCH** *(Pyrrhula pyrrhula)* Size of large Sparrow. Brightly coloured bird with short, strong beak and black cap. White wing band, white rump conspicuous in flight. ♂: bright red breast. ♀: mainly reddish grey. Juvenile: like ♀, minus black cap. Downy plumage blackish grey. Voice: call-note easily imitated, soft 'deu', song inconspicuous. Widespread nester in pine-woods, parks and, less often, gardens. In winter occurs in small parties by roadside and in gardens. Breeding season: IV—VII, 2 broods. Nest: shallow, made of twigs, rootlets and lichen, at moderate height in conifer, thicket, etc. Eggs: 4—5, blue with a few blackish brown spots, incubated 13 days by ♀. Young fed by ♂ and ♀. Diet: seeds and buds of woody plants. Partial migrant and resident bird.

Order: Passerines - Passeriformes **Family: Weavers** - Ploceidae

2 **SNOW FINCH** *(Montifringilla nivalis)* Size of large Sparrow. Grey head, brown back, black throat spot (not in winter and juvenile plumage). Wings mainly white, with black tips. Tail white, black down centre. In flight resembles Snow Bunting, but differentiated by grey head and black throat spot. Ground-dweller. Voice: call-note 'tsweek' or 'tswaa', etc. Song low twitter. Sporadic alpine bird, nests at high altitudes, occasionally roams about at lower altitudes. Breeding season: V—VII, 1—2 broods. Nest: litter of blades of grass, roots, moss, lichen and feathers, like Sparrow's, usually in rock crevice. Eggs: 5—6, pure white, incubated 18 days (and young afterwards fed) by ♂ and ♀. Diet: seeds, insects. Resident bird.

3 **ROCK SPARROW** *(Petronia petronia)* Small sparrow. Both sexes resemble female House Sparrow, but have white droplet spots at tip of tail and lemon yellow throat spot (absent in juvenile). Voice: call-note disyllabic, rather croaking 'pey-i', otherwise various hoarse, typical sparrow calls. Sporadic nester in warm, rocky, hilly country. Breeding season: V—VI, 2 broods a year. Nest and eggs: as for House Sparrow. Nest built in rock crevice, hole in wall, ruins, etc. Incubation period not known. Hatching period about 20 days; both parents feed young. Diet: seeds, berries, sometimes insects. Resident bird.

1

2

3

1 ♂
1 juv
1 ♂
1 ♀
2 juv
2 ♂
2 ♂
3 ♂
3 juv

1 **HOUSE SPARROW** *(Passer domesticus)* Familiar bird with distinct sex differences. ♂: grey crown, brown nape, black 'bib'. ♂ and juvenile: dull brown crown, pale throat. Downy plumage not developed. Voice: loud chirrups, when excited 'teuteuteuteu'. Very common nester in cultivated country, constant associate of man. Breeding season: IV—VIII, 3 broods a year. Several ♂♂ court 1 ♀ in striking joint display. Hops on ground. Nest: large pile of straw, blades of grass, paper; deep, feather-lined cup often roofed over. Built in hole in building, sometimes in open, on tree. Eggs: 5—6, white, sparsely or densely covered with blackish brown spots, incubated 14 days (and nestlings fed 17 days) by ♂ and ♀. Diet: seeds, grain, buds, fruit, occasionally insects. Resident bird.

2 **SPANISH SPARROW** *(Passer hispaniolensis)* Distinguished from very similar House Sparrow by chestnut crown, larger black 'bib' and vividly black-striped sides. ♀ and juvenile: resemble House Sparrow. Nests in colonies in bushy river valleys, in open woods and on trees lining country roads. Behaviour, care of brood and diet: as for House Sparrow. Resident and nomadic bird.

3 **TREE SPARROW** *(Passer montanus)* Size of House Sparrow. Brown crown, narrow white neck band, small 'bib' and black ear spot differentiate both sexes from House Sparrow and Spanish Sparrow. Juvenile: duller colouring. Downy plumage not developed. Voice: varied calls, e.g. 'chick, chee-ip, tchup, tak-tak-tak', etc., different from other most common species, i.e. House Sparrow. Common nester in cultivated country with sparse groups of trees, gardens and avenues. Breeding season: IV—VIII, 2—3 broods a year. Nest: in tree hole, nest-box or hole in wall, untidy mass of straw, blades of grass, roots and feathers. Eggs: 5—6, white, usually thickly brown-speckled, incubated 13—14 days (and young fed 14—16 days) by ♂ and ♀. Diet: as for House Sparrow. Partial migrant.

1

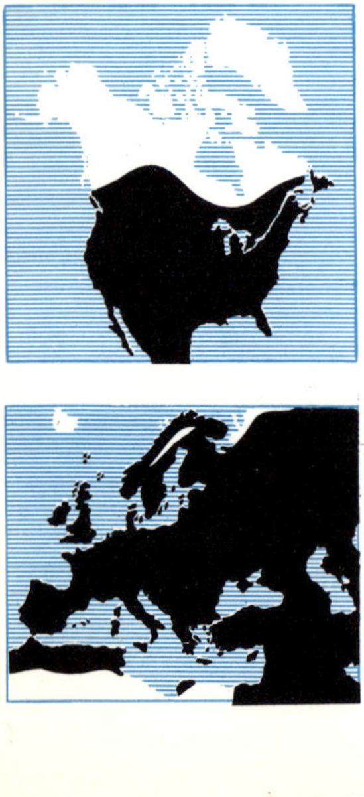

2

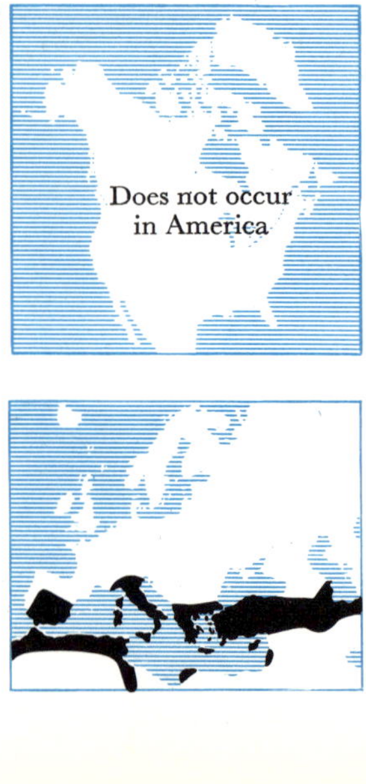

3

1 ♀
1 ♂ Italian Sparrow
1 ♂
2 ♀
2 ♂
3

1 **ROSE-COLOURED STARLING** *(Sturnus roseus)* Size of Starling. Pink body, black, crested head, black wings and tail. ♂ = ♀. Juvenile: sandy, with yellow beak. Voice and behaviour: as Starling, also consorts in flocks during breeding season. Nests in dry, grassy steppes. Breeding season: V—VII, 1 brood a year. Nest: made of twigs, blades of grass, roots and feathers, in hole in wall or ground, pile of stones, etc. Eggs: 5—6, blue, unspotted, incubated 11—13 days by ♀. Nestlings fed 14—19 days by ♂ and ♀. Diet: insects, especially locusts, also fruit. Migrant and irruption bird.

2 **STARLING** *(Sturnus vulgaris)* Somewhat smaller than Blackbird, with short tail. Largely black, with metallic lustre. ♂ = ♀. Juvenile: umber, whitish throat. Downy plumage greyish white. Flies fast in straight line, with whirring wing strokes. Voice: calls 'spatt' and 'tcheerr'. Song whistling and explosive, usually accompanied by wing-flapping. Common nester in open country with gardens, meadows, fields and ponds. Breeding season: IV—VII, 1—2 broods. Nest: made of grass, straw, twigs and feathers in hole in tree or wall or nesting-box. Often nests gregariously. Eggs: 5—6, greenish blue, unspotted, incubated 14 days (and nestlings fed about 20 days) by ♂ and ♀. After end of breeding season forms large flocks numbering thousands of birds, which roost in company in reeds, trees or on buildings. Diet: insects, worms, slugs, snails. Partial migrant. Migration: II—III and X.

3 **SPOTLESS STARLING** *(Sturnus unicolor)* Resembles Starling, but all black, no speckling. ♂ = ♀. Juvenile: darker than young Starling. Behaviour and voice: like Starling. Usually nests gregariously in villages, ruins, rocks and old hollow trees. Outside breeding season lives in flocks, often together with Starling. Breeding season: IV—VI, 2 broods a year. Nesting habits: as for Starling. Resident and nomadic bird.

1

2

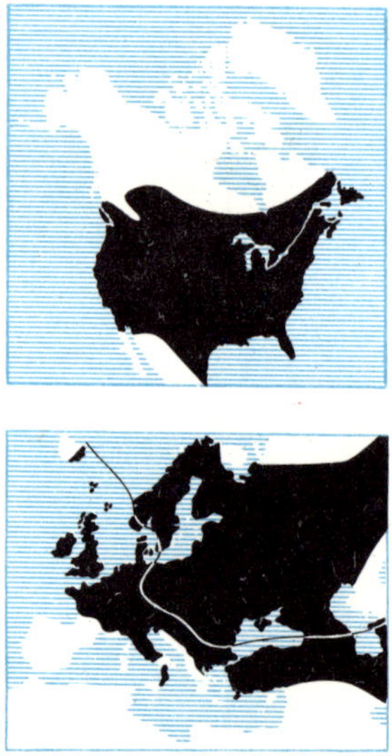

3

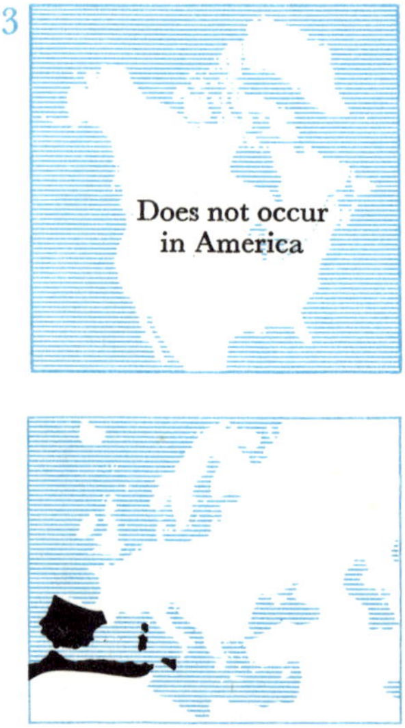

1 juv
1
2
2 ♂ S
2 juv
3
2 ♂ W

Order: Passerines - Passeriformes **Family: Orioles** - Oriolidae

1 **GOLDEN ORIOLE** *(Oriolus oriolus)* Size of Blackbird. ♂: vivid golden yellow and black. ♀: olive green back, light grey, dark-streaked underside. Juvenile: white, faintly streaked underside, light-edged wing feathers. Downy plumage brownish white. Lives retiringly in densest foliage, flies in flat curves. Voice: clear, loud, fluty 'weela-weeō', raucous 'krāāk'. Nests in groves and riverside woods in warm plains and hills, less often in pine-woods. Breeding season: V—VII, 1 brood a year. Nest: grass blades, phloem fibres, moss, wool and feathers, cleverly interwoven, hung below fork near tip of branch at top of tree. Eggs: 4, white, with a few brownish black spots, incubated 14 days (and nestlings fed about 2 weeks) by ♂ and ♀. Diet: insects, fruit. Migrant. Migration: V and VIII.

Order: Passerines - Passeriformes **Family: Crows** - Corvidae

2 **SIBERIAN JAY** *(Perisoreus infaustus)* Larger than Blackbird. Mainly greyish brown, with brown cap. Edge of wings, rump and outer tail feathers russet. ♂ = ♀. Juvenile: head feathers shorter and lighter shade of brown. Voice: loud 'kook, kook' and 'chair' calls. Lively, confident bird. Nests in northern conifer and birch forests. Breeding season: IV—V, 1 brood a year. Nest: made of brushwood, lined with lichen and feathers, on conifer, usually beside trunk. Eggs: 3—4, greenish white with greyish brown spots, incubated 16—17 days. Nestlings fed up to 5 weeks by ♂ and ♀. Diet: insects, birds' eggs, conifer seeds, berries. Resident and nomadic bird.

3 **JAY** *(Garrulus glandarius)* About size of pigeon. Variegated, reddish brown, with black tail, white rump, black moustachial stripe, shimmering blue on wings. ♂ = ♀. Juvenile: darker, less vividly coloured. Downy plumage not developed. Frequently ruffles feathers on crown. Flight ungainly. Voice: call-note loud, hoarse 'skraaak', often utters mewing 'pee-oo'. Common forest bird. Outside breeding season may occur in troupes, which mainly frequent oak stands. Breeding season: IV—V, 1 brood. Nest: flat, made of brushwood, stems, roots and blades, hidden among foliage. Eggs: 5—6, greenish, covered with small brown spots, incubated 16—17 days by ♂ and ♀. Hatching period: 19—20 days. Diet: acorns, beech- and hazel-nuts, berries, fruit, insects, worms, birds' eggs, mice. Partial migrant.

1

2

3

1 ♂
1 ♀
2
3

1 **NUTCRACKER** *(Nucifraga caryocatactes)* Size of small pigeon. Dark brown, with small white droplet spots. Under tail-coverts and end of tail white. ♂ = ♀. Juvenile: lighter brown, with whitish throat. Laborious, flapping flight (like Jay). Voice: distinctive loud, rattling 'krair'. Nests in extensive conifer forests, especially in high mountains; in some years abundant in lowlands as irruption bird. Breeding season: III—IV, 1 brood a year. Nest: made of brushwood, lichen, moss and blades of grass, high up in conifer, usually beside trunk. Eggs: 3—4, whitish, with a few brown spots, incubated about 18 days by ♀. Nestlings fed 23—25 days by ♂ and ♀. Diet: hazel-, cedar- and beech-nuts, conifer seeds, berries, fruit, insects, snails, etc. Resident and nomadic (also irruptive) bird. Main migration time: X—XI.

2 **AZURE-WINGED MAGPIE** *(Cyanopica cyanus)* About size of pigeon, with long, gradated tail. Pitch black head, azure wings and tail, black beak and legs. ♂ = ♀. Juvenile: duller, head more greyish black. Consorts in conspicuous, noisy parties. Voice: plaintive 'zhreee' and chattering 'klikkiklikkiklikki'. Breeding season: V—VI, 1 brood. Nest: made of dry twigs, softly lined with grass blades, wool and hair, in tree. Eggs: 5—9, olive brown, with large brownish black spots. Length of incubation and hatching period not known. Diet: as for Magpie. Resident bird.

3 **MAGPIE - BLACK-BILLED MAGPIE** *(Pica pica)* Size of pigeon, with long, gradated tail and gleaming black and white plumage. With vivid markings, 'paddling' wing strokes and long tail, unmistakable in flight. ♂ = ♀. Juvenile: shorter tail, black parts of plumage without gloss. Downy plumage absent. Occurs mostly in small parties. Voice: loud, chattering, rarely somewhat muted. Widespread nester in open, cultivated country with trees. Breeding season: IV—V, 1 brood a year. Nest: large, mostly arched over, made of brushwood, lined with blades of grass and hair, usually in lower branches of tree or in bush. Eggs: 6—7, greenish, thickly marked with dark brown spots and speckles, incubated 17—18 days by ♀. Nestlings fed 22—27 days by ♂ and ♀. Diet: insects, small mammals, birds, birds' eggs, snails, fruit, berries, crops. Resident bird.

1

2

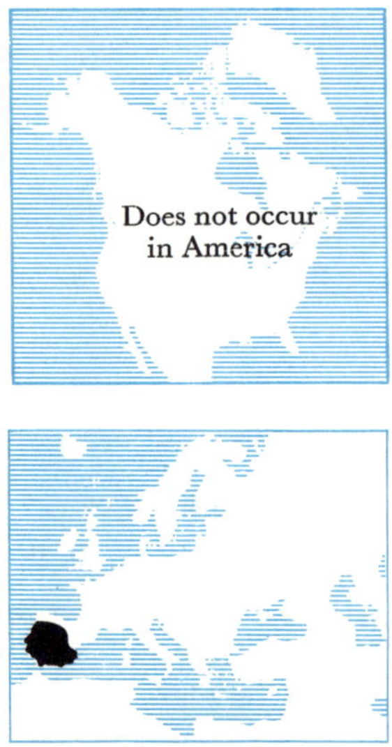

3

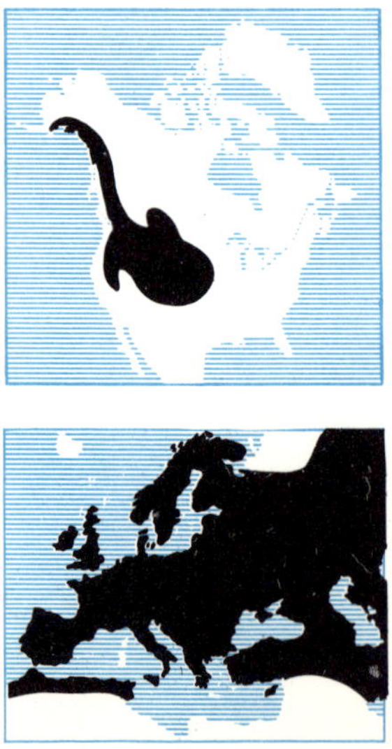

1
2
3

1 **CHOUGH** *(Pyrrhocorax pyrrhocorax)* Size of pigeon. Glossy bluish black, with curved red beak, red legs. ♂ = ♀. Juvenile: dull, black, beak and legs brownish yellow. Usually lives in large groups. Flies with skill, often glides, avoids trees, mostly perches on rocks or treeless ground. Voice: Jackdaw-like 'chough' and 'kyew' calls. Rock-dweller, inhabits mountains and cliffs. Breeding season: IV—V, 1 brood a year. Nest: pile of brushwood, lined with blades of grass and hair, in hollow or crevice, also in ruins. Eggs: 3—5, yellow or greenish, with fine brown speckles, incubated 21 days by ♀. Nestlings fed up to 40 days by ♂ and ♀. Diet: worms, snails, insects, berries, fruit. Resident bird.

2 **ALPINE CHOUGH** *(Pyrrhocorax graculus)* Size of pigeon. Uniform glossy black, with yellow beak, red legs. ♂ = ♀. Juvenile: dull black, legs and beak brownish. Voice: call-note clear, metallic 'skree', otherwise short 'tchiup' call. Breeding season: IV—VI, 1 brood. Nests gregariously on bare rocks in alpine zone, in winter lives near human communities. Nest: made of twigs and blades of grass, lined with hair and feathers, in crevice, hole or hollow. Eggs: 4—5, whitish, thickly covered with brown spots, incubated 18—21 days by ♀. Nestlings fed 31—38 days by ♂ and ♀. Diet: insects, worms, snails, small vertebrates, berries, fruit. Resident bird.

3 **JACKDAW** *(Corvus monedula)* Size of pigeon. Black, nape and sides of neck slate grey. White eyes. ♂ = ♀. Juvenile: generally duller. Downy plumage: sparse grey down feathers. Voice: regularly audible call-note clear 'tchak' or 'kyow', warning call screeching 'karr-r-r'. Common, gregarious nester in cultivated country with old trees, isolated quarries, in ruins and on church towers. Breeding season: IV—VI, 1 brood. Nest: made of brushwood and straw, lined with paper, hair and feathers, in tree hole, rock crevice, hole in wall or loft. Eggs: 4—6, greenish blue or grey, brown-speckled, incubated 17—18 days by ♀. Nestlings fed about 4 weeks by ♂ and ♀. Diet: insects, worms, snails, slugs, small vertebrates, berries, cereals. Partial migrant.

1

2

3

1
1 juv
2
3

1 **ROOK** *(Corvus frugilegus)* Typical crow size. Black, with violet blue lustre. Light, unfeathered skin at base of beak. ♂ = ♀. Juvenile: base of beak feathered. Distinguished from very similar Carrion Crow by thinner beak and long femoral feathers ('trousers'). Downy plumage dark grey, sparse. Voice: deep, short 'kaw' or protracted 'crow', sometimes high-pitched 'kirrr'. Common bird in flat country with tilled fields and groups of trees. Nests in colonies. Breeding season: III—IV, 1 brood. Nest: rather large, made of twigs, at top of deciduous tree, usually with several others. Eggs: 3—5, greenish with brown spots, incubated 16—18 days by ♀. Hatching period about 30 days. Diet: insects, worms, snails, slugs, small vertebrates, cereals, berries, fruit, etc. Partial migrant. Migration: X and III.

2 **CARRION CROW** *(Corvus corone)* Typical crow size. Glossy black, strong beak. ♂ = ♀ = juvenile. Represented in central and eastern Europe by black and light grey Hooded Crow *(Corvus corone cornix)*. Hybrids found in contact zone. Voice: cawing, 'kraa-kraa-kraa'. Common nester in cultivated country with trees. Breeding season: IV—V, 1 brood a year. Nest: made of twigs and brushwood, lined with turf, moss and hair, usually high up in tree. Solitary nester. Eggs: 5, bluish green with brownish black spots, incubated 17—18 days by ♀. Hatching period about 1 month. Diet: small vertebrates, birds' eggs, insects, worms, carrion, cereals, fruit, berries. Resident bird and partial migrant.

3 **RAVEN - COMMON RAVEN** *(Corvus corax)* About size of buzzard. All black. Massive beak, ruffled throat feathers. Distinguished in flight by triangular end of tail. ♂ = ♀ = juvenile. Voice: sonorous 'prruk, tok' calls, etc. Scattered bird, nesting mainly in rocky mountains, on steep cliffs and in a few lowland forests. In winter often found in large parties. Breeding season: III—IV, 1 brood a year. Nest: large, made of twigs and brushwood, with small, softly lined cup, high up in tree or rock niche. Eggs: 4—6, greenish blue, with brownish black spots and speckles, incubated 20—21 days by ♀. Nestlings fed about 40 days by ♂ and ♀. Diet: insects, vertebrates, carrion, fruit, refuse. Resident and nomadic bird.

1

2
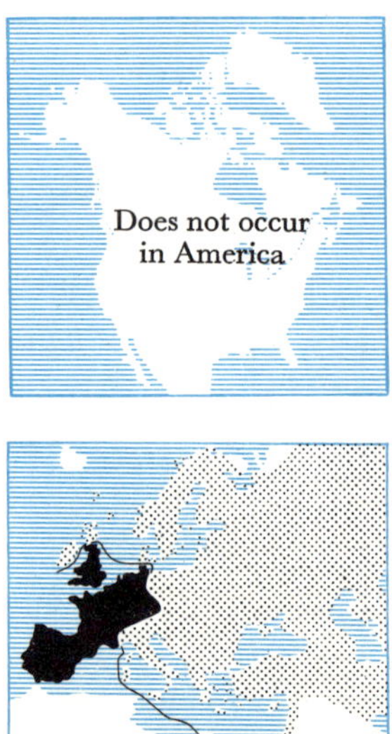

3
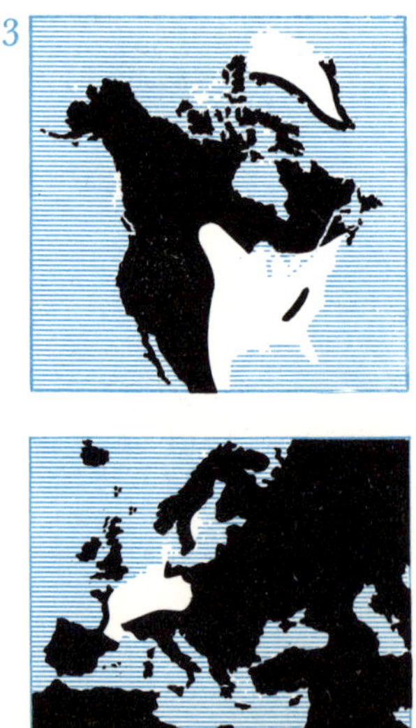

1 ad
1
1 juv
2
2 Carrion Crow
2 Hooded Crow
2 hybrid
3

PICTURES OF BIRDS IN FLIGHT

Flying ducks **Dabbling-ducks (1—5)**

Remain on surface of water, dive only if alarmed and for very short time. Sit high keeping tail clear of water. Dabble for food in shallow water and among aquatic plants. Rise easily, do not take off along surface. Show vividly coloured speculum on posterior border of wings when flying. ♂ and ♀ quite different in nuptial plumage, similar in winter plumage. Mates usually fly together, simplifying identification. Species more difficult to differentiate in full winter plumage; attention should be paid to distribution of colouring, shape of bill, build and, if possible, vocalization. In flocks composed of several species, differences in size are a useful indicator.

1 **Mallard** *(Anas platyrhynchos)*. Both sexes have metallic blue specula with white anterior and posterior border. ♂: dark green head with white collar, brown breast-plate. ♀: brown, longitudinally spotted, light-sided tail; specula diagnostic feature.

2 **Gadwall** *(Anas strepera)*. Both sexes have clearly visible white specula with brown anterior and black lateral border. ♂: grey-brown with white belly. ♀: somewhat slimmer than Mallard, wings more pointed, shorter beak.

3 **Pintail** *(Anas acuta)*. Both sexes distinguished by slim body, longish, thin neck and tapering, pointed tail. Metallic brown specula with white posterior border. ♂: dark head, white breast and belly, long 'pin' tail feathers. ♀: brown, longitudinally spotted, no light border on tail.

4 **Teal - Green-winged Teal** *(Anas crecca)*. Small, about half size of Mallard. Dark green specula with yellowish white anterior border. ♂: dark head, underside of tail yellow with black band, belly silvery white. ♀: dark brown with white belly. Small size diagnostic feature.

5 **Garganey** *(Anas querquedula)*. Small, light brown, specula light green with white anterior and posterior border. ♂: white eye stripe, anterior part of wings bluish grey, white belly sharply demarcated from light brown breast. ♀: light brown, anterior part of wings light grey.

1 ♂ S
1 ♀
3 ♂ S
2 ♀
3 ♀
2 ♂ S
5 ♀
4 ♂ S
4 ♀
5 ♂ S

1 Shoveler *(Anas clypeata)*. Smaller than Mallard, large, wide-tipped bill unmistakable. Green specula. ♂: bottle-green head, white breast, sharply outlined reddish brown belly, anterior part of wings bluish grey. ♀: brown, longitudinally spotted; light grey anterior part of wings and shape of bill diagnostic feature.

2 Wigeon - European Widgeon *(Anas penelope)*. Stocky, somewhat short neck, short, thin beak, poorly visible green specula. ♂: russet head, claret breast, large white patches on upper surface of wings. ♀: cinnamon, anterior part of wings grey, belly white.

Diving-ducks (3—5)

Sturdy, with rather short, wide body. Short tail usually lies on surface of water. Dive a great deal, looking for food on bottom in deep water; do not dabble. Have no metallic specula, but wings often marked with white bands and patches. Take off along surface, beating water with wings and legs.

3 Red-crested Pochard *(Netta rufina)*. Size of Mallard, broad white areas on upper surface of wings. ♂: variegated, with coral bill; russet head, grey back, black breast, broad black belly stripe. ♀: light greyish brown, wings largely white.

4 Pochard *(Aythya ferina)*. Both sexes have grey back and light grey wing bands. ♂: rust brown head, black breastplate, greyish white belly. ♀: greyish brown, no markings, light underside.

5 Ferruginous Duck *(Aythya nyroca)*. Little difference between sexes, both more or less uniform chestnut with pure white wing bands. ♂: head and breast reddish chestnut. ♀: duller (mahogany brown).

1 ♂ S
1 ♀
2 ♀
2 ♂ S
3 ♂ S
3 ♀
4 ♂ S
4 ♀
5

1 **Tufted Duck** *(Aythya fuligula)*. Both sexes have black back, white belly and sharply contrasting white wing bands. ♂: black back and breast, brilliant white sides, white belly. ♀: brownish black, with white underside.

2 **Scaup - Greater Scaup** *(Aythya marila)*. Generally resembles Tufted Duck, but ♂ has grey back. ♀ dark brown, with white ring at base of bill.

3 **Goldeneye - Common Goldeneye** *(Bucephala clangula)*. Both sexes distinguished by robust body and large white areas on wings. ♂: vivid black and white, sharply defined white cheek spots, black dorsal stripe. ♀ brown head with white neck-band, grey back.

4 **Eider - Common Eider** *(Somateria mollissima)*. Large, with short, thick neck and wedge-shaped head. ♂: back white, belly black, wings white with black primaries and secondaries, transitional plumage very chequered. ♀: completely dark brown.

5 **Long-tailed Duck - Oldsquaw** *(Clangula hyemalis)*. Small, with short bill. Both sexes have uniformly dark wings and often white head. ♂: dark brown wings and almost black, narrow, anteriorly forked dorsal stripe, body otherwise white. ♀: dark back and wings, head mainly white, with dark cheek spots.

1 ♀
2 ♀
1 ♂ S
2 ♂ S
3 ♂ S
3 ♀
4 ♂ S
4 ♂ S
4 ♀
5 ♂ S
5 ♀

1 Velvet Scoter - White-winged Scoter *(Melanitta fusca)*. Large and thickset. Both sexes black except for striking white specula. ♂: black, with sharply defined white area on primaries. ♀: blackish brown with 1—2 indistinctly defined, light spots on sides of head; brilliant white specula.

2 Common Scoter *(Melanitta nigra)*. Entirely dark (mostly black), without markings on wings. ♂: completely black, without white specula. ♀: blackish brown, sides of head and belly grey, without white specula.

Mergansers (3—5)

Slim ducks with rather thin neck, ± developed crest and thin, hook-tipped bill. Sit deep in water, dive frequently, remaining long time under water, catch fish. Fly quickly, stiffly outstretched in horizontal attitude.

3 Goosander - Common Merganser *(Mergus merganser)*. Larger, and in particular longer, than Mallard, with thin, red bill. ♂: greenish black head, wide black band down back, rest of body mostly white, underside often salmon pink. ♀: grey, with russet crest and specula on upper surface of wings.

4 Red-breasted Merganser *(Mergus serrator)*. Almost size of Mallard. Similar colouring to Goosander, but with two distinct dark stripes in specula. Long, thin, red bill. ♂: greenish black head, broad rust brown breast band, grey sides. ♀: rust brown colouring of head blends gradually with brown-grey colouring of back.

5 Smew *(Mergus albellus)*. Between Mallard and Teal in size. Short, dark bill. Largely white, ♀ with brownish red cap. ♂: mostly brilliant white, black lore spots, black back. ♀: back grey, underside white, russet cap.

1 ♂ S
1 ♀
2 ♂ S
2 ♀
3 ♂ S
3 ♀
4 ♂ S
4 ♀
5 ♂ S
5 ♀

Mostly large swimming birds, with long neck and short legs. Sit high in water, with end of tail raised. Belly and under tail-coverts white. Sexes not differently coloured. Gregarious birds, always look for green food on land, but also dabble. Parents share in rearing of young. Large flocks usually fly in echelon formation.

1 Sheldduck *(Tadorna tadorna)*. Vividly marked plumage. Head and neck greenish black, body white and black, broad russet breast- and shoulder band, red bill. Large white area on wings and white back separated on both sides by longitudinal white bar. Black stripe down belly. ♂: protuberance on bill, wide belly stripe, metallic lustre on head. ♀: no protuberance on bill, duller colouring, narrower belly stripe.

2 Greylag Goose *(Anser anser)*. Large, mainly grey, silver grey areas on anterior border of wings, bill orange yellow to pink, with no dark markings. Vocal expression as domestic goose. ♂ and ♀ differ slightly only in size.

3 White-fronted Goose *(Anser albifrons)*. Smaller and generally darker than Greylag Goose. Adults distinguished by pure white forehead and black cross-stripes on belly. Bill like Greylag Goose's. ♂ = ♀. Juveniles distinguished from Greylag Goose by darker neck and dark upper surface of wings. Characteristic high, disyllabic calls.

4 Lesser White-fronted Goose *(Anser erythropus)*. Smaller than White-fronted Goose, with short bill and longer white 'blaze'. Rather pointed wings and rapid wing strokes reminiscent of duck. ♂ = ♀. Juvenile lacks white forehead.

5 Bean Goose *(Anser fabalis)*. About size of Greylag Goose, but browner and darker; in particular, whole upper surface of wings is dark. Underside unmarked. Bill black, with yellowish red, annular band. ♂ = ♀ = juvenile. Different vocal expression from Greylag Goose.

1
2
3
4
5

1 **Brent Goose - Brant Goose** *(Branta bernicla)*. Small, largely black, white tail, narrow white neck bar. Short, black bill. ♂ = ♀. Juvenile has no neck bar. Voice: deep, throaty 'rruk'.

2 **Barnacle Goose** *(Branta leucopsis)*. Larger than Brent Goose, white head, stripes along upper surface of wings, white belly. Small, black bill, pointed wings. ♂ = ♀. Juveniles have brownish black (instead of black) back and ash grey sides.

Swans (3—5)

Large, completely white swimming birds with long neck. No sex differences. Juveniles have grey-brown plumage. Sit high in water, submerge head and neck when looking for food, also dabble; sometimes graze on land. Fly with neck extended and slow wing strokes, usually in line or in V-formation.

3 **Whooper Swan** *(Cygnus cygnus)*. Extremely long neck. Bicoloured beak (black tip, sulphur yellow base) without protuberance. Adult plumage snow-white. ♂ = ♀. Juvenile greyish white, with dull red beak. Flies silently.

4 **Mute Swan** *(Cygnus olor)*. Adults snow-white. Orange red beak with black protuberance in front of forehead. ♂ = ♀. Juvenile greyish brown, with smooth, lead grey beak. Slow wing strokes produce rhythmical sound like 'kraoo kraoo kraoo'.

5 **Bewick's Swan - Whistling Swan** *(Cygnus bewickii)*. Somewhat smaller and more robust than Whooper Swan, with shorter, thicker neck. Beak smooth, mainly black, yellow at base, adults pure white. Juvenile silvery grey, with dull pink beak. Flies without any noticeable sound.

1
2
2
3
4
5

Large, eagle-like birds of prey, with long, broad wings, usually short, straight-ended tail and unfeathered head and neck. Powerful beak. Ruffed neck. Skilled gliders. Head retracted, hardly visible, during flight. No sex differences. Carrion-eaters.

1 **Griffon Vulture** *(Gyps fulvus)*. Long, broad wings with spread primaries. Wing-coverts sandy, wing feathers dark brown, making wings appear bicoloured. Tail very short and straight-ended. White head and neck retracted into white ruff, which can be seen above origin of wings. Glides in large circles without beating wings.

2 **Black Vulture** *(Aegypius monachus)*. Similar in size and appearance to Griffon Vulture, but uniform blackish brown. Head larger and rather lighter. Dark ruff visible in front of shoulders. Short, slightly triangular tail.

Eagles (3—5)

Large, powerful birds of prey, with long, and usually fairly wide, wings. Primaries spread during flight. Skilled gliders. Plumage generally brown, often with white patches or spots. Sexes often differ in size, but usually not in colouring. Juvenile different from adult plumage as a rule. Large nests (eyries) used several years. Prey: vertebrates of all groups.

3 **White-tailed Eagle** *(Haliaeëtus albicilla)*. Adult plumage dark brown, with yellowish brown head and neck. Long, wide wings, spread primaries. Large, sulphur yellow beak. Unfeathered tarsi. White, triangular tail. Juvenile more rust brown, with dark brown head and neck, often appears mottled; beak and tail brownish black. Complete adult plumage takes several years to grow, so that most half-grown juveniles have half-white tail and yellow-brown beak.

4 **Golden Eagle** *(Aquila chrysaëtos)*. Adult plumage uniform dark brown, with white at base of tail; head and nape golden yellow. Distinguished from White-tailed Eagle by slimmer body, longer, dark, rounded tail and black beak. Five spread primaries seen at tips of wings. Juvenile dark brown, with large white patch at base of wing feathers on underside of wings. Proximal half of tail white, broad dark band at end.

5 **Imperial Eagle** *(Aquila heliaca)*. Moderately large, strong eagle. Adult plumage dark brown, with light yellow crown and nape. A few gleaming white scapulars. Tail rather shorter than Golden Eagle's, without any white. Wing tips show 7 spread primaries. Juvenile plumage with honey-coloured to dark brown spots, rust brown crown, black brown wing feathers, no white scapulars.

1
2
3
3 juv
4
4 juv
5

1 **Spotted Eagle** *(Aquila clanga)*. Adult: moderately large, blackish brown eagle with broad wings showing 7 spread primaries. Upper tail-coverts more or less white. Short, rounded tail. Projecting, relatively small head. Juvenile: coverts form large white spots on upper surface of wings, join in 2—3 white lines when wings are spread. White base of tail especially clear.

2 **Lesser Spotted Eagle** *(Aquila pomarina)*. Smaller species resembling Spotted Eagle. Differences: in flight, wings and tail look smaller, 6 spread primaries visible at wing tips, base of tail black (rarely with indistinct white spots). Often has light brown spot on nape. Juvenile: yellowish white spots on upper surface of wings form several light lines when wings are spread, but less vivid than in Spotted Eagle. Back of neck rusty yellow.

3 **Booted Eagle** *(Hieraaëtus pennatus)*. Adult: size of buzzard, with fairly long, rectangular tail. Quicker wing strokes than buzzard, flight more graceful, flies among tops of trees. Two colour variants. Light variant: creamy yellow back, with dark wings and scapulars, underside yellowish white (including tail), wings dark. Dark variant: entirely dark brown, only tail a little lighter. Juvenile: narrow streaks on white to rust-coloured underside distinctive.

4 **Short-toed Eagle** *(Circaëtus gallicus)*. Larger than buzzard. Round, owl-like head, yellow eyes. Eagle-like flight, often hovers. Adult: underside almost white, with a few rust brown cross-marks. Neck darker, primaries black-tipped. Rather long tail with 3—4 dark cross-bands. Back greyish brown. Juvenile: underside brown, with dark spots.

5 **Osprey** *(Pandion haliaëtus)*. Larger than buzzard. Long, angular wings, white underside, wide, dark eye stripe. Underside of wings mainly white, with black spot on carpal joints. Narrow black bands on tail. Feathers on back of head often erected. Frequently hovers over water, swoops, seizing fish in talons. Juvenile distinguished by light-edged feathers on back.

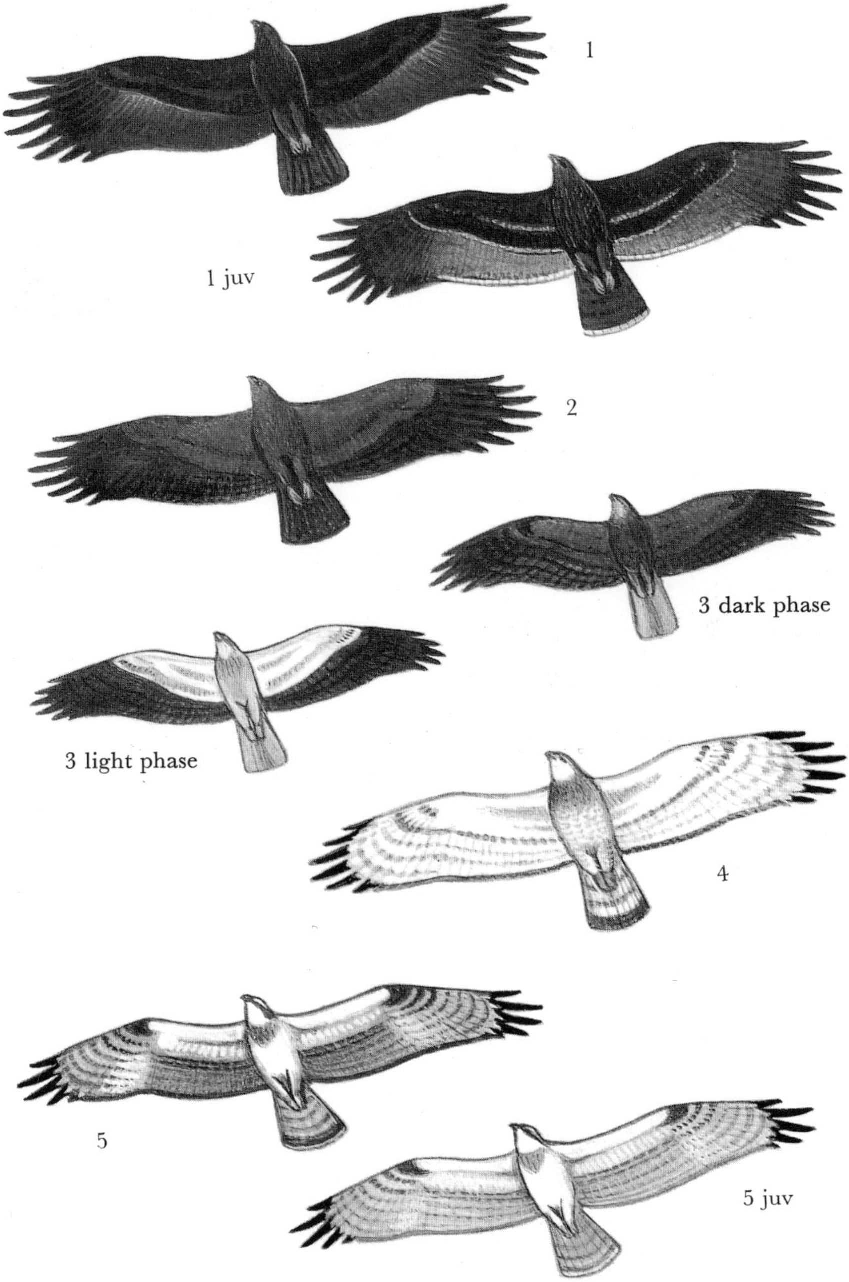
1
1 juv
2
3 dark phase
3 light phase
4
5
5 juv

Moderately large, rather ungainly, with broad wings and wide, rounded tail. Rounded head with small beak usually protruding little. Often circle and glide high in air.

1 Buzzard *(Buteo buteo)*. Larger than crow. Wide wings showing 5 short, spread primaries. Tail mostly fanned, with at least 6 narrow bands. Often glides in circles, dark carpal joint spots on under side of spread wings only faintly visible. Several colour variants; commonest are dark brown, all intermediate shades from russet to those with an almost completely white underside. Juveniles to some extent distinguishable by longitudinal, rather than transverse markings on underside.

2 Rough-legged Buzzard - Rough-legged Hawk *(Buteo lagopus)*. Distinguished in flight from similar Common Buzzard by generally paler head, dark belly patch on almost white underside, distinct dark carpal joint spots on underside of wings, intensely black primaries and single broad band at end of very pale tail. Frequently hovers close to ground, showing legs feathered down to toes. Juveniles hardly distinguishable from adults in open.

3 Long-legged Buzzard *(Buteo rufinus)*. Larger than Common Buzzard, often quite pale, russet to rust brown, with pale head. Tail cinnamon to white, with no distinct cross-bands and no terminal border (unlike all other buzzards). Juveniles indistinguishable from adults.

4 Honey Buzzard *(Pernis apivorus)*. Superficially resembles Common Buzzard, but shows following differences in flight. Wings somewhat narrower and longer, with several dark stripes on underside. Head not rounded, but tapers forwards, like pigeon's. Tail longish, with dark border at end and 2 cross-bands near base. Plumage very variably coloured, especially underside, from plain coffee to broad cross-bands, or almost white, dark-streaked specimens. Juvenile distinguished from grey-headed adults by often almost white head and spotted back.

1 DV

1 LV

2 DV

2 LV

3 DV

3 LV

4 DV

4 LV

Characterized by short, rounded wings, long tail and thickly cross-banded underside in adult plumage (hawks) and long, angular wings and forked tail (kites). All are skilled fliers. Hawks hunt mainly in tree-tops, kites in open country. Tree-nesters.

1 **Sparrow Hawk** *(Accipiter nisus)*. Small, no larger than pigeon. Short, wide wings, rapid wing strokes. After a few quick strokes, glides with spread wings. Long, straight-ended tail with 4 dark bands. Often surprises songbirds by flying low over ground. Thin, yellow, long-toed legs. ♂: back bluish grey, underside white with dense, rust brown ripple marks. ♀: about one third larger, with grey-brown back and thick grey-brown ripple marks on underside. Juvenile: brown back, underside markings less fine.

2 **Goshawk** *(Accipiter gentilis)*. Size of Buzzard. ♂ about one third smaller than ♀. Resembles Sparrow Hawk in flight and in hunting methods, but mostly catches larger birds. Often shows white supraorbital stripe and long, white under tail-coverts. Yellow eyes. ♂: back greyish to slate grey, underside densely ripple-marked. ♀: only difference size. Juvenile: back browner, large, dark brown spots down underside.

3 **Black Kite** *(Milvus migrans)*. Size of Buzzard. Dark brown, with lighter head and neck, underside russet. In flight, no distinct wing markings, small notch in tail hardly noticeable when tail is fanned. Light, graceful flight. ♂ = ♀. Juvenile distinguished by somewhat darker head.

4 **Red Kite** *(Milvus milvus)*. Size of large buzzard, with slim, angular wings. 5 primaries visible when wings are spread. Distinguished by fairly long, deeply forked tail. Head and tail very light russet, large, whitish areas on underside of wings, near tip. Often glides in circles. ♂ = ♀.

Top of p. 267: Flight silhouettes of different types of birds of prey. Attention should be paid to differences in shape of wings and length of tail.

Golden Eagle

Common Buzzard

Marsh Harrier

Goshawk

Peregrine Falcon

Slim birds of prey, with long, rather narrow wings and long, slightly rounded tail. Flutter, usually fairly close to ground. After a few wing strokes, often glide for a short time, holding wings in characteristic V-position.

1 **Marsh Harrier** *(Circus aeruginosus)*. Size of Buzzard, mostly dark brown, with slightly rounded wings; lives near water. ♂: variegated, blue-green area on wings, light grey tail, rusty yellow head and neck. ♀: plain dark brown, plain-coloured wings. Head (except sides) and carpal joint yellowish white. Juvenile: like ♀, but head and neck darker (more rust brown).

2 **Hen Harrier - Marsh Hawk** *(Circus cyaneus)*. Smaller than buzzard. White rump, distinct facial disc. ♂: light grey; black-tipped wings with narrow, black posterior border. ♀: brown back, light brown, striped head, light rusty yellow, streaked underside, 4 dark bands on tail. Juvenile: similar to ♀, but with darker marks on underside.

3 **Montagu's Harrier** *(Circus pygargus)*. Smallest harrier, slimly built. ♂: ash grey with black-tipped wings and narrow black band above secondaries; russet-streaked underside. ♀: brown back, small white rump spot; whitish, russet-streaked underside, banded tail. Juvenile: like ♀, but underside plain russet.

4 **Pallid Harrier** *(Circus macrourus)*. Closely resembles 2 and 3. ♂: light grey back, white underside, black-tipped wings; no white rump (cf. 2) and no black wing bands (cf. 3). ♀: dark brown back, narrow white rump: underside light brownish grey, with dark longitudinal markings, broad cross-bands on tail. Juvenile: like ♀, but with rusty yellow, unspotted underside.

1 ♂
1 ♀
2 ♂
2 ♀
3 ♂
3 ♀
4 ♂

Small to moderately large, powerfully built birds of prey, with rather long, pointed, usually narrow wings and often triangularly tapering tail. Majority chase and catch prey on wing, only a few species hover. When striking, wings are held close to body at different angles, so that they appear sickle-shaped, or give body form of torpedo, substantially increasing velocity of attack.

1 **Peregrine Falcon** *(Falco peregrinus)*. Dark falcon, about size of crow. Fast, pigeon-like flight, flat, but powerful, wing strokes. ♂: slate grey back, wide, black moustachial stripe, close black cross-bands on underside; throat and breast white. ♀: much larger and mostly darker than ♂, underside usually russet-tinted. Juvenile: rusty yellow brown, moustachial stripe generally narrow; back dark brown, underside with vivid dark stripes.

2 **Saker Falcon** *(Falco cherrug)*. Almost same size as Peregrine Falcon, but somewhat slimmer and longer-tailed in flight, with wider wings. Head whitish, with indistinct moustachial stripe, wings rust brown. Underside, including breast, thickly brown-spotted. ♂ = ♀ (♀ may have more thickly spotted under side). Juvenile: back generally darker, with light-edged feathers; broad brown stripes on crown.

3 **Hobby** *(Falco subbuteo)*. About size of Kestrel, with blackish grey back, large moustachial stripe, thickly streaked underside, russet under tail-coverts and legs; long, curved wings, short, tapering tail (often resembles swift in flight). ♂ = ♀. Juvenile: rusty yellow-edged feathers make back appear browner; top of head lighter, under tail-coverts duller.

4 **Merlin - Pigeon Hawk** *(Falco columbarius)*. Smallest European falcon, slightly larger than Blackbird. Flies rapidly, mostly close to ground. Flight interspersed with brief gliding periods, sometimes also flutters like swallow. ♂: back bluish grey, thickly striped, rusty yellow underside; bluish grey tail with wide black band at end. No distinct moustachial stripe. ♀: back grey-brown, underside whitish, with brown stripes, cross-banded tail. Juvenile = ♀.

5 **Kestrel** *(Falco tinnunculus)*. Size of Collared Turtle Dove, mostly rust brown. Frequently hovers with fanned tail. ♂: rust brown, spotted back, ash grey head and tail, wide black tail band. ♀: rust brown, cross-banded back, head and tail. Juvenile: resembles ♀, but colouring generally duller.

6 **Red-footed Falcon** *(Falco vespertinus)*. Small falcon, resembles Hobby in flight. Mostly gregarious, often hunts flying insects after dusk. Frequently hovers. ♀: plain slate grey, appears almost black; legs and under tail-coverts brick-red; feet, cere and orbital ring red. ♀: cinnamon head and underside, grey back with black cross-bands. Juvenile: brownish back, striped underside.

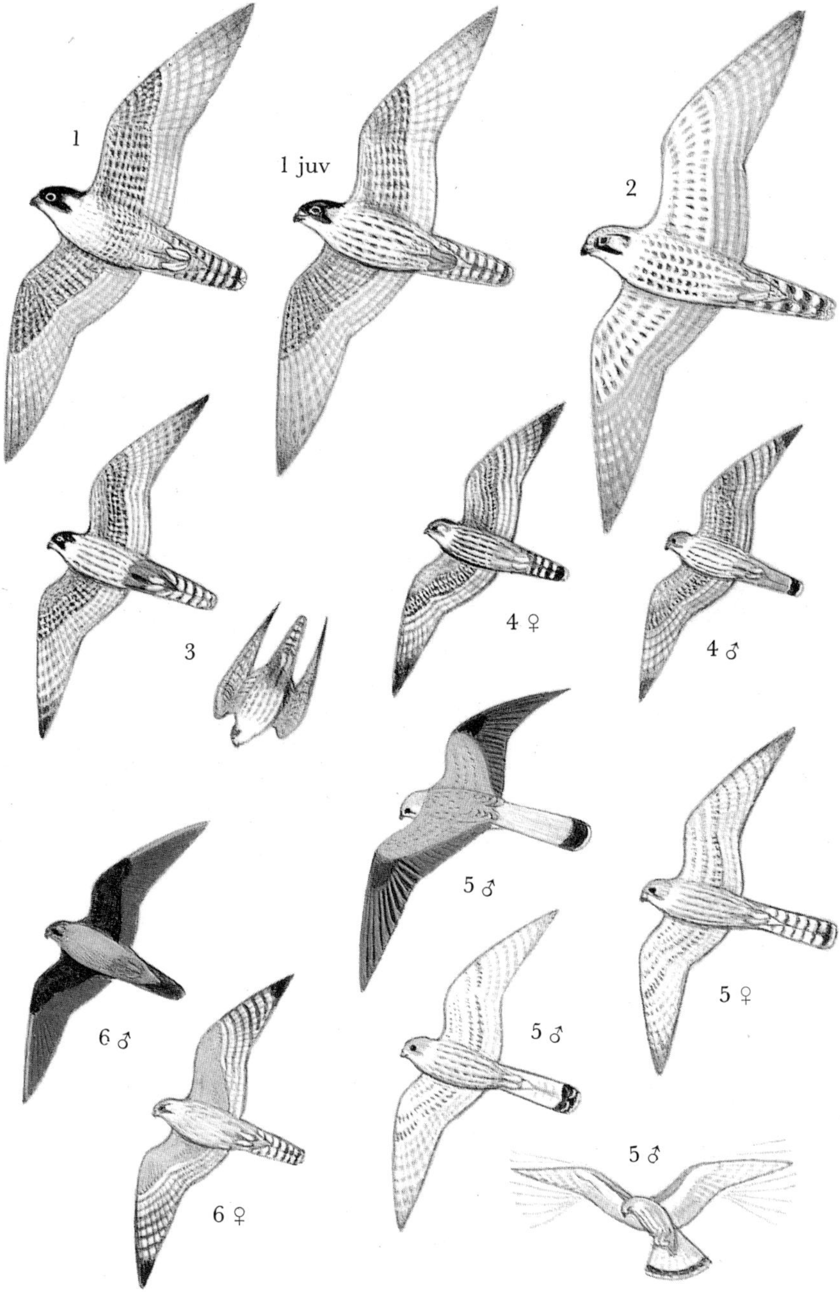
1
1 juv
2
3
4 ♀
4 ♂
5 ♂
5 ♀
6 ♂
5 ♂
6 ♀
5 ♂

Large, black and white, littoral birds, with long, blunt-tipped, flat-sided, bright red beak and red legs. Forage for food in sand and mud. Rest with beak tucked in dorsal plumage. Fly with flat wing strokes.

1 Oystercatcher - American Oystercatcher *(Haematopusost ralegus)* Size of domestic pigeon. Head and anterior part of wings black, white areas on wings, posterior part of back and tail white, with sharply defined, black border. Underside pure white. Vermilion legs and beak. ♂ = ♀.

Plovers and Lapwings (2—3)

Moderately large and small waders, with stocky body, short neck, large head and large eyes. Short, pigeon-like beak. Mostly very sociable littoral birds with peculiar form of locomotion (run short distance with stiff body, then suddenly stop dead). With sandy back and head markings, are excellently adapted to sandy environment and gravel-banks.

2 Lapwing *(Vanellus vanellus)*. Size of domestic pigeon. Wide, rounded wings, slow, strenuous wing strokes, metallic green dorsal plumage, narrow crest, broad-, white-based tail with black border. Black neck band (in ♀ greyer).

3 Turnstone *(Arenaria interpres)*. Size of thrush. Short beak, chequered plumage with large amount of white. Wings have rust brown anterior border and stripe. Black U-stripe, open in front, on white rump, black band at end of tail. ♀ less vividly marked. Winter plumage grey on back.

Avocets and Stilts (4—5)

Large species with long legs and neck, black and white plumage and long, thin beak.

4 Avocet *(Recurvirostra avosetta)*. Size of domestic pigeon, mainly white. Top of head, tips of wings and 4 wide stripes on back and upper surface of wings deep black. Thin, upcurved beak. Long legs stretch far beyond end of pure white tail. ♂ = ♀.

5 Black-winged Stilt - Black-necked Stilt *(Himantopus himantopus)*. Size of pigeon. White with completely black wings and black back. Thin, straight beak. Extremely long, stilt-like, red legs stretch beyond end of tail by almost body length.

1
2
3 S
3 W
4
5

1 **Ringed Plover** *(Charadrius hiaticula)*. Size of thrush. Sandy back, black, white and grey markings on head. Black, complete breast-band. White wing stripes (cf. Little Ringed Plover). Dark tail framed with white. Orange yellow feet. ♂ = ♀. Flies rapidly. Tuneful, rising display call, 'pee-i'.

2 **Little Ringed Plover** *(Charadrius dubius)*. Size of lark. Closely resembles Ringed Plover, but upper surface of wings plain sandy brown. White-sided tail. ♂ = ♀. Juvenile lacks variegated head design.

3 **Kentish Plover** *(Charadrius alexandrinus)*. Size of lark. Fawn back, white wing stripes, breast band incomplete (only black spot on either side of breast). Black legs. Head markings black in ♂, greyish brown in ♀.

4 **Grey Plover** *(Pluvialis squatarola)*. Size of Lapwing. Pointed wings. Stone grey back with white rump and whitish wing bands. Nuptial plumage black on underside, with black axillaries. Underside in winter plumage light, but axillaries black. ♂ = ♀. Flies rapidly, with regular wing strokes.

5 **Golden Plover** *(Pluvialis apricaria)*. About size of Laughing Dove. Golden brown back, black underside (nuptial plumage), or brownish breast and white belly (winter plumage). Axillaries white, wing band absent or very indistinct (cf. similar Grey Plover). ♂ = ♀.

1
2
3
4 W
4 W
4 S
5 W
5 S
5 W

1 Dotterel *(Eudromias morinellus)*. About size of thrush, with dark, brown-grey back and light supraorbital stripe. Dark breast with narrow white band, underside blackish brown. Winter and juvenile plumage light. Narrow white border on tail. Flies quickly, with heavy wing strokes.

Sandpipers (2—4)

Slim waders, mostly of moderate size, with small head, thin, straight beak longer than head and long legs which show beyond end of tail in flight. Fly fast, with pointed wings curved. Frequent margins of inland waters in parties, wade in shallow water in search of food. Many species nod head or jerk body. Sexes same colour. Characteristic vocal differences very useful identification guide (see text, pp. 100—104).

2 Spotted Redshank *(Tringa erythropus)*. Larger than Blackbird. Nuptial plumage slaty black, underside of wings light, legs and base of beak crimson. No wing band (cf. Redshank). Legs extend well beyond end of tail. Winter and juvenile plumage: lattice pattern on upper surface of grey-brown wings, rump white, tail with numerous grey cross-bands.

3 Redshank *(Tringa totanus)*. Size of Blackbird, brownish grey, with dark-spotted back, broad, white posterior wing border and light cross-bands on tail. Bright red feet project noticeably beyond end of tail. Winter and juvenile plumage lighter, not particularly spotted.

4 Marsh Sandpiper *(Tringa stagnatilis)*. Size of Starling. Back grey-brown, with fine, dark spots, underside largely white, white rump. Tail white, with faint grey markings in middle. Beak very thin, blackish. Olive grey legs extend well beyond end of tail. Winter and juvenile plumage grey on back; underside, rump and tail white.

1 S
1 W
2 S
2 W
2 S
3 S
4 S
4 W

1 **Greenshank** *(Tringa nebularia)*. Larger than Blackbird. Nuptial plumage brown, finely spotted back, wings darker, with no white markings, end of back and rump white, tail white, with a few cross-bands, beak long and black, green legs project well beyond end of tail. Winter and juvenile plumage very light (back grey-brown, underside white).

2 **Green Sandpiper** *(Tringa ochropus)*. Size of thrush, very dark (appear almost black), rump brilliant white, base of tail white. No light markings on upper surface of wings, under surface dark (cf. Wood Sandpiper). Legs relatively short, do not extend far beyond end of tail. Winter and juvenile plumage differ little from nuptial plumage.

3 **Wood Sandpiper** *(Tringa glareola)*. Size of thrush. Brown back, white rump not sharply defined, cross-banded tail, under surface of wings light (cf. Green Sandpiper). Legs relatively short, do not project far beyond end of tail. Winter and juvenile plumage similar.

4 **Common Sandpiper - Spotted Sandpiper** *(Tringa hypoleucos)*. Size of lark. Fairly short beak, short legs hardly show behind end of tail. Back olive brown, distinct white wing bands, white-sided tail. Breast (especially in nuptial plumage) grey, belly white. Peculiar mode of flight, with flat, jerky wing strokes, close to surface of water. ♂ = ♀.

1 S
1 W
2
2
3
3
4
4

Small to moderately large waders, usually with small head, thin beak about same length as head and comparatively short legs. Mostly very gregarious. Frequent coasts when migrating, often in huge, dense flocks with remarkably coordinated flight. Run about with tripping gait near water. Sexes similar, summer and winter plumage usually different. Vocal expression not very striking as a rule.

1 Knot *(Calidris canutus)*. Size of Blackbird, recognizable by bulky shape. Nuptial plumage: underside russet, under surface of wings light, back rust brown, rump and tail-coverts light-spotted. Winter plumage: back ash grey, dark brown spots on breast, rest of underside white. Juvenile's back more brown.

2 Little Stint *(Calidris minuta)*. Barely size of Sparrow, with short beak. Back largely rust brown, vividly marked with black spots and light, V-shaped stripes. Outer tail feathers light grey. Underside white, except for grey-brown sides of crop. Winter and juvenile plumage grey, with pure white underside.

3 Temminck's Stint *(Calidris temminckii)*. Size of small Sparrow. Plain, mousy grey back. Blurred grey sides of crop stand out against white underside. Outer tail feathers white (cf. similar Little Stint). Winter plumage more single shade of grey. Juvenile plumage more olive brown, with light-edged feathers, on back.

4 Dunlin *(Calidris alpina)*. Size of Starling. Rust brown back, dark grey wings with light under surface, spotted crop, black belly patch. Middle of rump and tail dark, end of tail brownish grey. Winter plumage completely light on underside. Juvenile plumage blackish brown on back, with brown-edged feathers.

5 Curlew Sandpiper *(Calidris ferruginea)*. Size of Starling, with upcurved beak. Nuptial plumage rust brown, with russet underside. Rump white, underside of wings light. Winter plumage: back grey-brown, underside white. Juvenile plumage: back grey-brown, breast has rusty yellow tinge, rest of underside white.

1 S
1 W
2 S
2 W
3 S
3 W
4 S
4 W
5 S
5 W

1 **Sanderling** *(Calidris alba)*. Size of Starling, with rather short beak and short, shiny black legs. Nuptial plumage rust brown, with pure white belly. Broad white wing bands, anterior border of wings black. Winter plumage light grey, underside pure white. Juvenile plumage: black-spotted back.

2 **Ruff** *(Philomachus pugnax)*. ♂ (Ruff) in nuptial plumage appears thick-necked, owing to folded ear-tufts and bright ruff. Often looks very brightly coloured. Indistinct band on upper surface of wings. Legs frequently yellowish red. ♀ (Reeve) almost ⅓ smaller than ♂, in nuptial plumage has brown, dark-spotted back; rump and upper tail-coverts white on sides only. Winter and juvenile plumage like ♀.

Phalaropes (3—4)

Resemble sandpipers in build, but adapted more to water. Often swim far from side, sitting high in water. Have natatory membranes on toes. Nod head. Sexes differ; in association with reverse care of brood (♂ incubates eggs and leads young), females are larger and more brightly coloured.

3 **Grey Phalarope - Red Phalarope** *(Phalaropus fulicarius)*. Size of small thrush. Nuptial plumage: white cheeks, black crown, rust brown, black-spotted back, bright rust brown underside, white wing bands; tail and middle of rump dark, beak and feet yellowish. ♂ less brightly coloured. Winter and juvenile plumage very light; light grey back with black nuchal spot and eye stripe, underside pure white. White band and black posterior border on upper surface of wings.

4 **Red-necked Phalarope - Northern Phalarope** *(Phalaropus lobatus)*. Size of lark. Nuptial plumage: head and back almost black, white chin, broad russet neck-band, underside white, beak thin and black. Winter and juvenile plumage: back grey and black, darker (blacker) than in Grey Phalarope, especially in middle of back; white wing bands, pure white underside.

Stone Curlews (5)

Fairly large, ungainly waders with thick beak and legs. Large head with strikingly large, yellow eyes. Inhabit barren country and are not usually bound to water. Twilight birds.

5 **Stone Curlew** *(Burhinus oedicnemus)*. Size of Wood Pigeon. Sandy brown, with rows of dark spots. Two sharply defined, white bands on upper surface of wings. Flies with slow wing strokes. ♂ = ♀. Juvenile more rust-coloured on back.

1 S
1 W
2 ♀
2 ♂ W
2 ♂ S
3 S
3 W
5
4 W
4 S

Large, dark brown sea-birds with hooked beak. Adults have long middle tail feathers. Similar form and mode of flight to gulls. ♂ and ♀ indistinguishable. Bird of prey habits; seize young and eggs of other sea-birds, also snatch prey from latter in violent flying attacks. Often settle on water.

1 **Arctic Skua - Parasitic Jaeger** *(Stercorarius parasiticus)*. Size of Black-headed Gull. Either dark brown with lighter underside (dark variant) or white on underside, with sharply defined smoky black crown (light variant). White shafts of primaries form white area on wings (especially clear at close quarters). Adults: middle pair of tail feathers tapering, several cm longer than others. Juvenile: middle tail feathers short.

2 **Pomarine Skua - Pomarine Jaeger** *(Stercorarius pomarinus)*. Size of Common Gull, closely resembles Arctic Skua, but adults have spatulate, twisted middle tail feathers. White area on wings also seems more distinct. Two colour variants (dark- and light-bellied) and intermediate forms exist.

3 **Long-tailed Skua - Long-tailed Jaeger** *(Stercorarius longicaudus)*. Smallest skua, barely size of Black-headed Gull, slimmer than Arctic Skua, with lighter breast, more sharply contrasted black cap and greyer back. Adults best identified by very long, pointed, flexible middle tail feathers. White area on wings poorly developed. Juvenile: browner, with short tail feathers; hard to distinguish from young Arctic Skua in flight.

4 **Great Skua - Skua** *(Stercorarius skua)*. Size of Common Gull, ungainly, dark brown, with white-shafted primaries on rather wide wings. End of tail triangular, beak hooked, beak and feet black. Juvenile almost indistinguishable in flight. Flies unwieldily, distinguished from other skuas by large, ungainly body and short tail.

1 LV
1 DV
1 imm
2 LV
2 DV
3
4
3 imm

Mostly large sea-birds with long wings. Adult plumage mainly pure white; only colouring bluish grey or slate grey to black 'mantle' on back and upper surface of wings. Some species, in nuptial plumage, wear dark brown to black cap, replaced in winter by white feathers. Conversely, many large gulls have fine, dark streaks on head and neck in winter plumage. Juveniles have more heavily pigmented plumage (mostly various shades of brown). Beak hook-tipped. ♂ and ♀ indistinguishable. Skilled fliers. Flesh-eating.

1 **Glaucous Gull** *(Larus hyperboreus)*. Size of Great Black-backed Gull. Very pale (at 2 years entirely white, older birds have light grey mantle). Wings not black-tipped. Beak yellow with red spot at tip. Legs pink. Juvenile completely light brown, later cream; dark beak.

2 **Great Black-backed Gull** *(Larus marinus)*. Larger than Herring Gull. Large yellow beak, red-tipped lower mandible. Legs flesh pink. Mantle black; narrow white border on anterior and posterior edge of black upper surface of wings, small white wing tips. Juvenile: blackish brown-spotted back, underside at first dark-spotted, later white. Flies with slow, heavy wing strokes, rather like heron, sometimes appears unwieldy.

3 **Lesser Black-backed Gull** *(Larus fuscus)*. Size of Herring Gull. Adults coloured similarly to Great Black-backed Gull, but have thinner beak. Legs yellow. English populations have slate grey mantle instead of black. Juvenile: covered with dark brown spots, hard to distinguish from Herring Gull; wings appear generally blacker, middle of back also darker (more blackish grey); black tail band more sharply separated from white tail coverts; beak black, legs dull pink. Flies with greater agility than 2.

4 **Herring Gull** *(Larus argentatus)*. Large gull with bluish grey mantle, yellow beak, red spot at tip of lower mandible, flesh pink feet. Winter plumage: dark streaks on head and neck. Juvenile: brown, with dark brown primaries and brown band at end of tail (cf. 3). Beak black, feet flesh pink.

5 **Common Gull - Mew Gull** *(Larus canus)*. Resembles Herring Gull, but smaller and slimmer. Rather small, greenish yellow beak and feet, no red spot on beak. Wing tips black and white. Juvenile: back covered with brown spots, underside with faint brown spots, later white; bicoloured beak (dark tip), bluish pink legs, sharply defined black band at end of tail. Wing strokes faster than in Herring Gull, steadier than in Black-headed Gull.

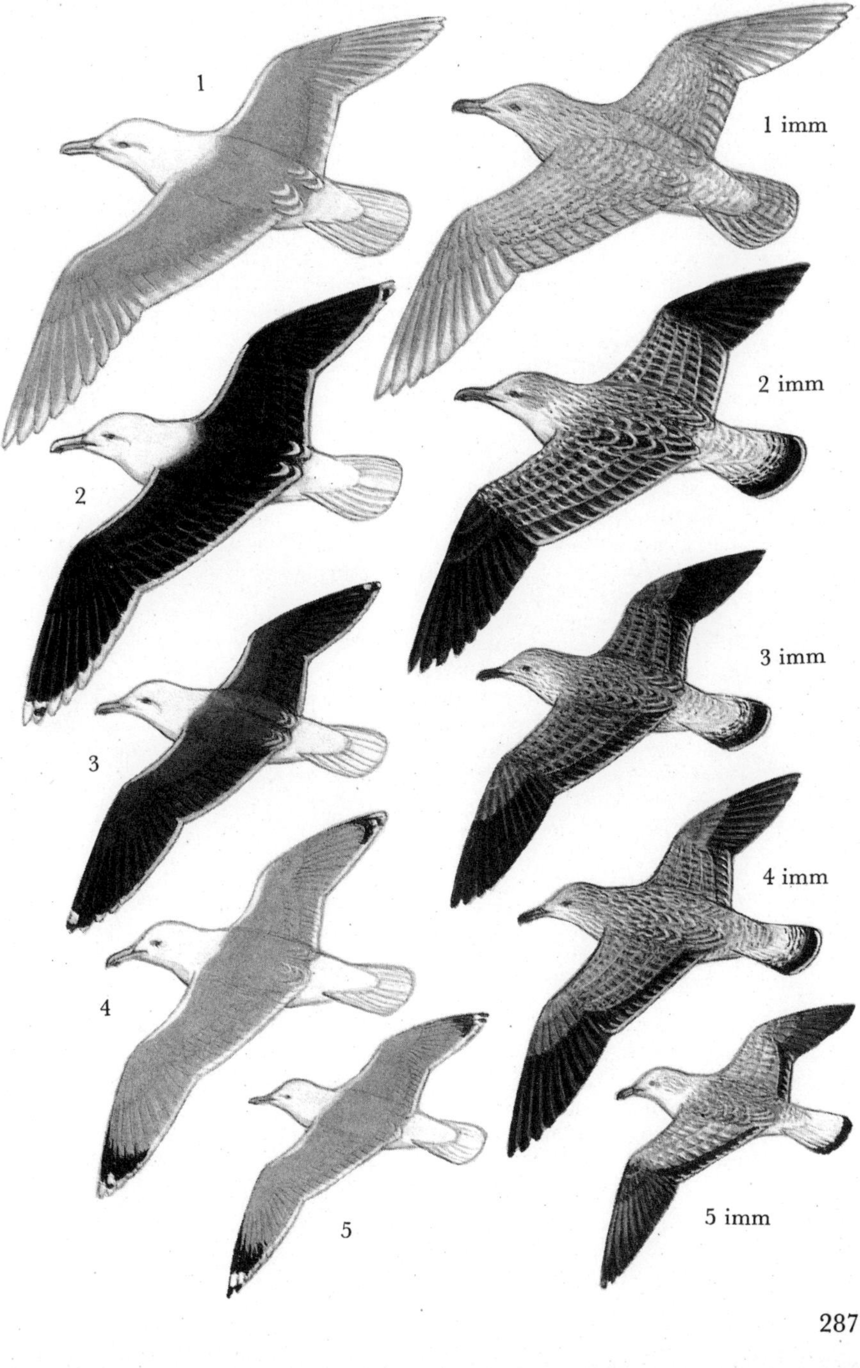
1
1 imm
2 imm
2
3 imm
3
4 imm
4
5
5 imm

1 Kittiwake - Black-legged Kittiwake *(Rissa tridactyla)*. Little larger than Black-headed Gull. Grey mantle, black-tipped wings (no white), yellow beak, black legs. Straight-ended or slightly notched tail. Winter plumage: grey stripes on sides of head. Juvenile: grey mantle, black nuchal stripe, black band at end of tail, black M-shaped band across wings.

2 Black-headed Gull *(Larus ridibundus)*. Coffee-coloured hood, light grey mantle. Anterior border of wings white, narrow black posterior border. Legs and beak dark red. Winter plumage: white head with dark ear-spot, legs and beak light red. Juvenile: fawn to white crown, brown wing-coverts, black band at end of tail, legs and beak brownish yellow.

3 Mediterranean Gull *(Larus melanocephalus)*. Size of Black-headed Gull. Pitch black hood, large red beak, wings without black tips. Winter plumage: white head with dark stripes round eyes and on nape. Juvenile: brown wing-coverts, black band at end of tail, black primaries (in Black-headed Gull white).

4 Little Gull *(Larus minutus)*. Smaller than Black-headed Gull, with somewhat rounded wings and fluttering flight. Head pitch black to middle of neck. Wing tips white (no black markings), under surface of wings dark. Winter plumage: crown and nape blackish grey, black ear-spot, light grey back. Juvenile: crown and nape blackish grey, tail white with black band at end, black M-shaped stripe across upper surface of wings.

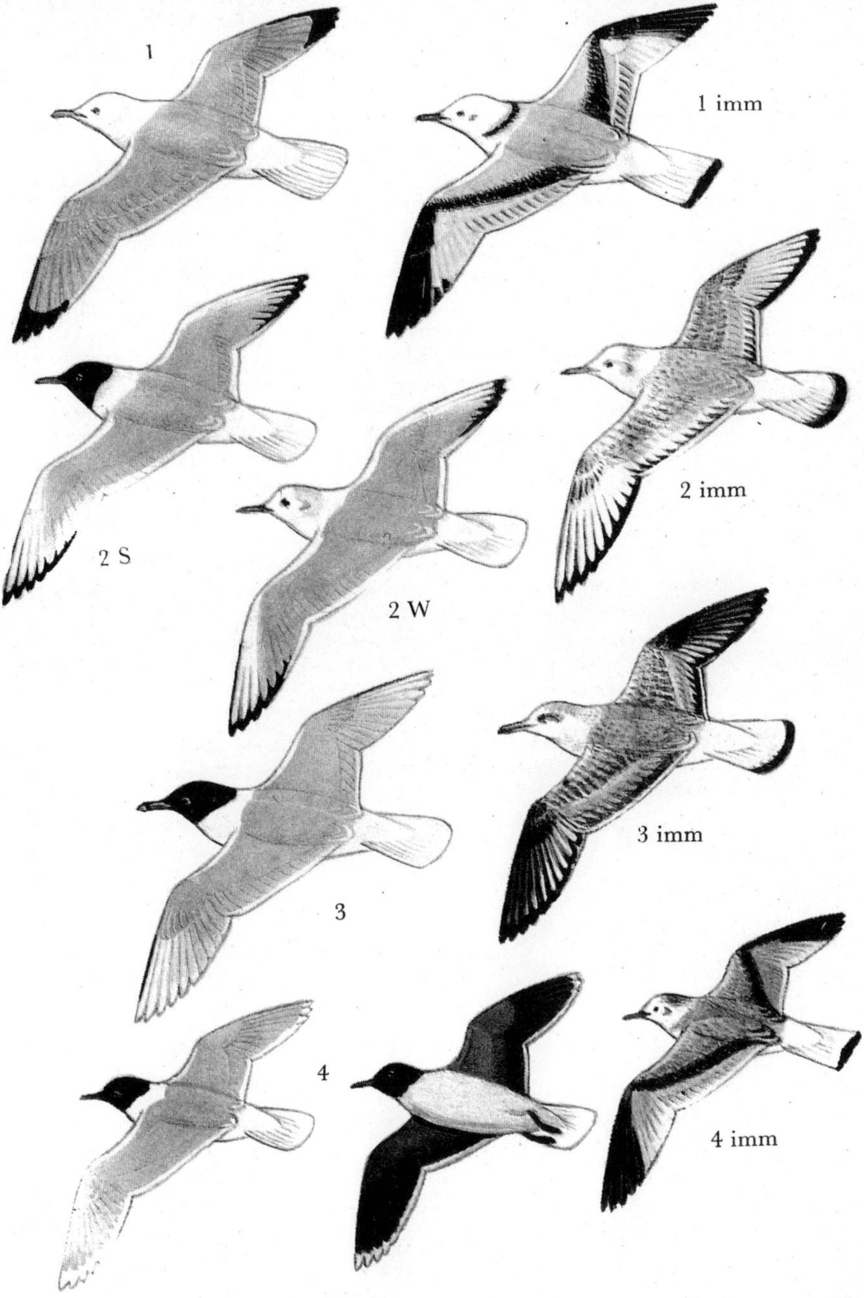
1
1 imm
2 S
2 W
2 imm
3
3 imm
4
4 imm

Slender, narrow-winged sea-birds, mostly with light grey back and black crown. Tail more or less deeply forked. Beak usually thin and evenly tapered. Legs short. ♂ and ♀ identical. Nuptial and winter plumage almost same. Juvenile plumage distinguished by partly dark brown back. Fly skilfully, mostly with slow wing strokes. Many species hover, watching for prey, then plunge vertically into water with folded wings. Majority eat fish, a few inland species live mainly on insects. Unlike gulls, seldom settle on water. Sociable birds, often form large colonies with closely packed nests.

1 **Caspian Tern** *(Hydroprogne tschegrava)*. Size of Common Gull. Largest tern. Large, powerful, bright red beak, longish crown feathers on black cap. Fairly wide wings, underside of primaries darker. Tail only slightly forked. Flies like gull. Winter plumage: brown-spotted back, orange beak. Harsh, croaking voice, reminiscent of Grey Heron's, helps in identification.

2 **Gull-billed Tern** *(Gelochelidon nilotica)*. Size of Black-headed Gull. Black, gull-like beak, light grey tail only slightly forked. Flies like gull, rarely dives to water. Winter plumage: minus black cap, black only behind eyes, nape grey. Juvenile plumage: head as in winter plumage, but crown and nape often cream; brownish black spots on back, beak and legs reddish brown.

3 **Sandwich Tern** *(Sterna sandvicensis)*. Size of Black-headed Gull. White, deeply forked tail. Relatively long, black, yellow-tipped beak. Black cap with long nape feathers. Winter plumage: cap intermingled with white feathers, forehead white. Juvenile plumage: brown to black spots on back, tail tips blackish brown.

4 **Little Tern - Least Tern** *(Sterna albifrons)*. About size of Swift, smallest tern. Moderately forked tail. Quick wing strokes. Regular hovering. Nuptial plumage: white forehead, with sharply, but unevenly defined black cap. Yellow, black-tipped beak. Winter plumage: head completely white, except for black nape. Juvenile plumage: white crown, dark nape, blackish anterior wing border.

1
1
2
2 imm
3
3 imm
4 imm
4

1 **Common Tern** *(Sterna hirundo)*. Slimmer and smaller than Black-headed Gull. Black cap, brick red, black-tipped beak, deeply forked tail. Easy, buoyant flight, hovers over surface and plunges vertically into water. Winter plumage: white forehead, dark secondary coverts on upper surface of wings. Juvenile plumage: crown rust-coloured and blackish brown, brown cross-bands on back.

2 **Arctic Tern** *(Sterna paradisaea)*. Same shape as Common Tern, but throat, breast and belly light grey and only sides of head, along edge of black cap, pure white. Extremely long, dark grey tail feathers. Coral beak and legs. Winter plumage: white forehead, grey back and white tail (somewhat longer than Common Tern's). Juvenile plumage: white forehead, grey (not brownish grey) back.

3 **Roseate Tern** *(Sterna dougallii)*. Same size as Common Tern, but looks lighter and has very deeply forked tail. Beak red only at base, otherwise black. Legs red. Underside often pink-tinted in spring. Strikingly buoyant flight. Nuptial plumage: completely black, sharply defined cap. Winter plumage: white forehead, black nape. Anterior border of wings lighter shade of grey than in Common Tern. Juvenile plumage: small black spots on light back.

1 S
1 imm
2 S
2 W
3 S
3 W

1 **Black Tern** *(Chlidonias niger)*. Smaller than Common Tern. Slate grey, head and underside black. Beak and legs reddish black. Under wing-coverts and rump grey (cf. White-winged Black Tern). Winter plumage: head black- and white-spotted, black scapular spot, underside white. Juvenile plumage: brown-speckled crown and back.

2 **White-winged Black Tern** *(Chlidonias leucopterus)*. Same size as Black Tern, vivid black and white markings. Head, belly and under surface of wings black. Tail and upper wing-coverts pure white (in much commoner Black Tern, under surface of wings is light and upper surface dark). Beak and legs red. Winter plumage: no black scapular spot; anterior border of wings less dark than in Black Tern. Juvenile plumage: upper surface of wings pale grey, back dark brown (darker than Common Tern).

3 **Whiskered Tern** *(Chlidonias hybrida)*. About same size as Black Tern. Looks very pale. Catches prey in plunging dive, like Common Tern. Nuptial plumage: black cap, light grey back, grey breast, black belly. Beak and legs blood red. Winter plumage: forehead white, nape grey to black, no black scapular spot (cf. Black Tern). Juvenile plumage: blackish brown cross-bands on back and upper surface of wings; forehead, sides of head and neck and belly white.

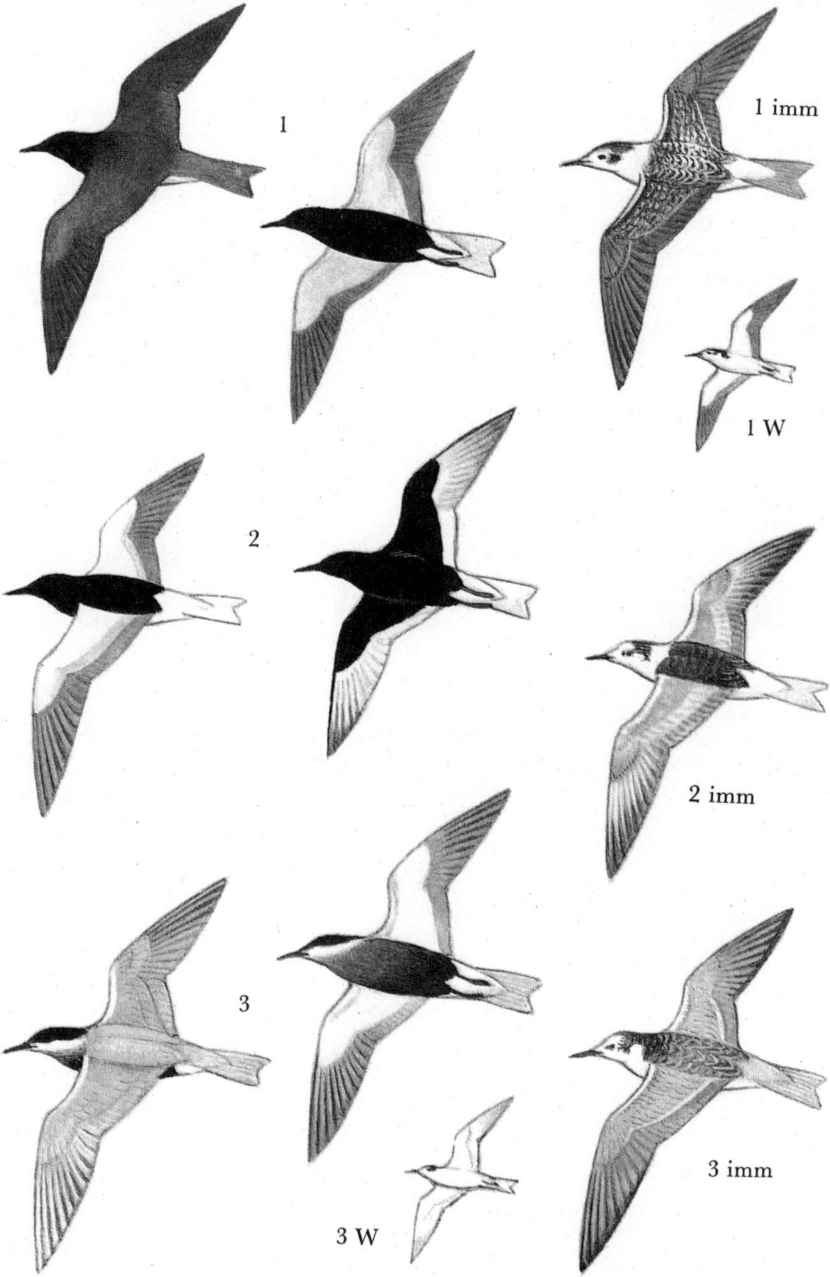
1
1 imm
1 W
2
2 imm
3
3 imm
3 W

PICTURES OF EGGS AND NESTS

(for details see p. **313—320***)*

Apart from a few exceptions, the colour plates show a complete clutch of each of the given species as it appears in the nest, i.e. the typical colouring, markings and shape of the eggs and the shape and characteristic structure of the nest.
Fig. 39 shows the oven-like nest of the Dipper without the eggs. In Fig. 53 we can see a typical Nuthatch nest made of thin slivers of bark in a nesting box. The Bullfinch nest in Fig. 60 contains an incomplete clutch of only 2 eggs.
In most cases the eggs are shown less than their normal size. There is likewise no relationship between their sizes in the individual figures. The average size of the eggs is therefore given below each figure in millimetres.

1. Great Crested Grebe *(Podiceps cristatus)* (55×37)
2. Mallard *(Anas platyrhynchos)* (56×40)
3. Little Bittern *(Ixobrychus minutus)* (35×26)
4. Pintail *(Anas acuta)* (54×38)

5. Mute Swan *(Cygnus olor)* (110×74)

7. Marsh Harrier *(Circus aeruginosus)* (48×38)

6. Kestrel *(Falco tinnunculus)* (38×31)

8. Partridge - Gray Partridge *(Perdix perdix)* (35×27)

9. Pheasant - Ring-necked Pheasant *(Phasianus colchicus)* (45×36)

11. Little Crake *(Porzana parva)* (30×22)

10. Moorhen - Common Gallinule *(Gallinula chloropus)* (42×30)

12. Coot *(Fulica atra)* (53×35)

13. Lapwing *(Vanellus vanellus)* (46×32)

14. Ringed Plover *(Charadrius hiaticula)* (34×24)

15. Turnstone - Ruddy Turnstone *(Arenaria interpres)* (41×29)

16. Snipe - Common Snipe *(Gallinago gallinago)* (39×28)

17. Woodcock *(Scolopax rusticola)* (44×33)

18. Black-tailed Godwit *(Limosa limosa)* (55×38)

19. Curlew *(Numenius arquata)* (67×47)

20. Redshank *(Tringa totanus)* (43×30)

21. Ruff *(Philomachus pugnax)* (43×31)

22. Arctic Skua - Parasitic Jaeger *(Stercorarius parasiticus)* (57×40)

23. Stone Curlew *(Burhinus oedicnemus)* (54×38)

24. Black-headed Gull *(Larus ridibundus)* (52×37)

25. Common Tern *(Sterna hirundo)* (41 × 30)

27 Razorbill *(Alca torda)* (73 × 47)

26. Black Guillemot *(Cepphus grylle)* (58 × 40)

28. Wood Pigeon *(Columba palumbus)* (40 × 29)

29. Long-eared Owl *(Asio otus)* (40× 32)

30. Skylark *(Alauda arvensis)* (23× 17)

31. Wryneck *(Jynx torquilla)* (20× 15)

32. Swallow - Barn Swallow *(Hirundo rustica)* (19× 13)

33. Blue-headed Wagtail - Yellow Wagtail *(Motacilla flava)* (19× 14)

34. White Wagtail *(Motacilla alba)* (20× 15)

35. Sand Martin - Bank Swallow *(Riparia riparia)* (17× 13)

36. Tree Pipit *(Anthus trivialis)* (20× 15)

37. Red-backed Shrike *(Lanius collurio)* (22× 17)

38. Lesser Whitethroat *(Sylvia curruca)* (17× 13)

39. Dipper *(Cinclus cinclus)* (26× 19)

40. Hedge Sparrow — Dunnock *(Prunella modularis)* (19× 14)

41. Great Reed Warbler *(Acrocephalus arundinaceus)* (23 × 16)

42. Wood Warbler *(Phylloscopus sibilatrix)* (16 × 12)

43. Icterine Warbler *(Hippolais icterina)* (18.5 × 13.5)

44. Chiffchaff *(Phylloscopus collybita)* (15 × 12)

45. Willow Warbler *(Phylloscopus trochilus)* (15×12)

46. Redstart *(Phoenicurus phoenicurus)* (19×14)

47. Collared Flycatcher *(Ficedula albicollis)* (17×13)

48. Black Redstart *(Phoenicurus ochruros)* (19×14)

49. Robin *(Erithacus rubecula)* (20×15)

50. Blackbird *(Turdus merula)* (30×21)

51. Fieldfare *(Turdus pilaris)* (28.5×21)

52. Great Tit *(Parus major)* (17×13)

53. Nuthatch *(Sitta europaea)* (19× 14.5)

54. Yellowhammer *(Emberiza citrinella)* (21.5× 16)

55. Corn Bunting *(Emberiza calandra)* (24× 17)

56. Hawfinch *(Coccothraustes coccothraustes)* (24× 17.5)

57. Chaffinch *(Fringilla coelebs)* (19 × 14.5)

59. Greenfinch *(Carduelis chloris)* (20 × 14.5)

58. Siskin *(Carduelis spinus)* (16 × 12)

60. Bullfinch *(Pyrrhula pyrrhula)* (20.5 × 15)

61. Tree Sparrow *(Passer montanus)* (19×14)

62. Magpie - Black-billed Magpie *(Pica pica)* (33×23)

63. Golden Oriole *(Oriolus oriolus)* (30×21)

64. Carrion Crow *(Corvus corone)* (41×29)

EGGS AND NESTS

1 Great Crested Grebe *(Podiceps cristatus)*

Nest made of parts of different aquatic plants piled on surface of water and usually only anchored to edge of reed belt. The 3—6 white (later brown) eggs are covered with nest material when bird leaves nest. Size of eggs 55×37 mm.

2 Mallard *(Anas platyrhynchos)*

Nest made of dry plant parts in depression under bush or tangled plants, mostly near water, sometimes in tree hole. Clutch: 7—11 brownish to bluish green eggs measuring 56×40 mm, surrounded by ring of dark down.

3 Little Bittern *(Ixobrychus minutus)*

Nest conical basket made of thin twigs and sedge, reed or rush-grass blades, built in sedge or bush over surface of water. Clutch: 5—6 chalky white eggs measuring 35×26 mm.

4 Pintail *(Anas acuta)*

Nest depression in ground, in grass or under bush, lined with dry blades of grass and leaves. Clutch: 7—11 yellowish eggs, usually elongate, measuring 54×38 mm.

5 Mute Swan *(Cygnus olor)*

Nest large pile of brushwood, sedge, reeds or seaweed, lined with down. Stands in shallow water, or on ground directly beside water. Clutch: 5—8 large, brownish yellow eggs measuring 110×74 mm.

6 Kestrel *(Falco tinnunculus)*

Nest made of twigs and brushwood, usually taken over from crow, wood pigeon or bird of prey. Mostly high up in tree, on rock or old masonry, without bedding. Clutch: 5—6 almost spherical, brown-spotted eggs, average size 38×31 mm.

7 Marsh Harrier *(Circus aeruginosus)*

Nest made of sedge, reeds and brushwood on bent sedge blades, well hidden in dense rushes, built close to surface of water. Clutch: 4—5 round, white eggs measuring about 48×38 mm.

8 Partridge - Gray Partridge *(Perdix perdix)*

Nest shallow depression in ground, well hidden under bush, in grass or in matted plants, thinly lined with straw and feathers. The 10—20 plain greenish grey to brownish eggs closely match the background. Size of eggs: 35×27 mm.

9 Pheasant - Ring-necked Pheasant *(Phasianus colchicus)*

Nest on ground, hidden in grass, in heather or under bush, usually at edge of wood or in copse, generally unlined. Clutch: 10—12 plain greyish green eggs measuring about 45×36 mm.

10 Moorhen - Common Gallinule *(Gallinula chloropus)*

Bowl-like nest made of sedge leaves and straw, usually bedded on brushwood, built over shallow water or on ground near water. Clutch: 6—10 yellow eggs with brown to black spots and speckles. Size of eggs: 42×30 mm.

11 Little Crake *(Porzana parva)*

High-sided nest made of sedge and reed-grass leaves, built over water. Clutch: mostly 6—8, greyish yellow eggs with faded reddish brown spots. Size of eggs: 30×22 mm.

12 Coot *(Fulica atra)*

Nest usually hidden in edge of reed-grass, less often in open, in shallow water. Made of broken reed-grass leaves and blades. Shallow cup contains 6—9 yellowish white, densely black-speckled eggs measuring 53×35 mm.

13 Lapwing *(Vanellus vanellus)*

Nest in shallow depression in ground, usually in wet meadow, poorly lined with grass blades and leaves. Clutch: 4 top-shaped, olive green eggs, thickly marked with brownish black spots, lie with pointed end facing inwards. Closely match surroundings. Size of eggs: 46×32 mm.

14 Ringed Plover *(Charadrius hiaticula)*

Nest shallow depression in sand, mostly lined with small pebbles and mussel shells. Clutch: 4 top-shaped, sandy eggs with blackish-brown spots. Size of eggs: about 34×24 mm.

15 Turnstone - Ruddy Turnstone *(Arenaria interpres)*

Nest shallow depression on rocky sea-shore, lined with dry plant parts. Clutch: 4 pear-shaped, greyish green eggs with grey and brown spots. Size of eggs: about 41×29 mm.

16 Snipe - Common Snipe *(Gallinago gallinago)*

Nest made of material from dense sedge-bank in wet meadow or swamp; shallow, thinly lined, built on ground. Clutch: 4 pear-shaped, olive green eggs with large dark spots, usually lie with pointed end facing inwards. Size of eggs: 39×28 mm.

17 Woodcock *(Scolopax rusticola)*

Shallow nest lined with dry leaves and moss, mostly built in woodland beside tree trunk. Clutch: 4 wide oval, buff eggs with reddish brown spots and speckles, closely match forest floor. Size of eggs: 44×33 mm.

18 Black-tailed Godwit *(Limosa limosa)*

Nest made of dry grass in shallow depression in ground in wet meadow. Clutch: 4 top-shaped, olive brown eggs with faded dark brown spots. Size of eggs: 55×38 mm.

19 Curlew *(Numenius arquata)*

Nest shallow, thinly lined depression in ground in wet meadow, moor or heath. Clutch: 4 top-shaped, olive green eggs with dark brown spots. Size of eggs: about 67×47 mm.

20 Redshank *(Tringa totanus)*

Nest straw-lined depression in tall grass, usually well hidden by growing blades of grass. Clutch: 4 top-shaped grey eggs with dark spots. Size of eggs: about 43×30 mm.

21 Ruff *(Philomachus pugnax)*

Female (Reeve) incubates eggs and rears young unaided. Nest shallow depression in damp meadow or moor, lined with a few grass blades. Clutch: 4 top-shaped, greyish green eggs with dark brown spots. Size of eggs: about 43×31 mm.

22 Arctic Skua - Parasitic Jaeger *(Stercorarius parasiticus)*

Often nests in colonies. The 2 brownish green, dark-spotted eggs are laid in a shallow depression in moss or grass. Size of eggs: about 57×40 mm.

23 Stone Curlew *(Burhinus oedicnemus)*

Nests in shallow depression in ground. Nest lined with a few stones and plant parts, contains 2 light grey eggs with blackish brown spots and streaks. Size of eggs: 54×38 mm.

24 Black-headed Gull *(Larus ridibundus)*

Nest pile of reed-grass and rush fragments, often lined with dry grass. Frequently stands in shallow water or on clump of sedge. Eggs: 3, mostly olive green with brown spots, but ground colour very variable. Average size of eggs: 52×37 mm.

25 Common Tern *(Sterna hirundo)*

Nest very shallow, thinly lined depression in ground, mostly in sand or pebbles near water. Clutch: 3 light green to olive green, dark-spotted eggs measuring 41×30 mm.

26 Black Guillemot *(Cepphus grylle)*

No nest built. Eggs laid in rock crevice or hole in ground, usually on bed of small stones. Clutch: 2 oval, light, dark-spotted eggs measuring about 58×40 mm.

27 Razorbill *(Alca torda)*

No nest built. Single white or greenish egg, with dark spots and scrolls, is laid on cliff, on bare rock. Size of eggs: 73×47 mm.

28 Wood Pigeon *(Columba palumbus)*

Nest flat brushwood dish 30—40 cm across, built in tree fairly high above ground. Clutch: 2 pure white eggs measuring about 40× 29 mm.

29 Long-eared Owl *(Asio otus)*

Nest made of twigs and brushwood, usually by other bird (bird of prey, Crow or Wood Pigeon), mostly in tree. Clutch: 4—7 short elliptical eggs, average size 40× 32 mm.

30 Skylark *(Alauda arvensis)*

Nest built in depression in ground in field or meadow. Loose structure made of grass blades and rootlets, lined with fine blades and often hair. Clutch: 4—5 greyish yellow, thickly brown-speckled eggs. Size of eggs: 23—17 mm.

31 Wryneck *(Jynx torquilla)*

Nest in tree hole, not excavated by bird itself (unlike woodpeckers), but often partly improved. Sometimes in nest-box. Clutch: 7—10 glossy white eggs measuring about 20× 15 mm.

32 Swallow - Barn Swallow *(Hirundo rustica)*

Dish-shaped nest made of lumps of clay stuck together with saliva, lined with straw and a few feathers. Built indoors, e.g. in stable, barn, porch. Clutch: 4—5 white eggs with brownish red spots measuring 19× 13 mm.

33 Blue-headed Wagtail - Yellow Wagtail *(Motacilla flava)*

Nest usually well hidden in small depression in ground, among tufts of grass, often on slope, e.g. railway embankment, etc. Made of dry grass blades and rootlets, with smooth, hair-lined cup. Clutch: 4—6 whitish eggs thickly marked with cloudy, greyish brown spots. Size of eggs: about 19× 14 mm.

34 White Wagtail *(Motacilla alba)*

Nest built mostly in recess in rock, under roof or in half-open nest-box. Made of dry blades of grass, leaves, moss and rootlets, softly lined with hair and feathers. Clutch: 5—6 whitish, dark grey- and brown-speckled eggs. Size of eggs: 20× 15 mm.

35 Sand Martin - Bank Swallow *(Riparia riparia)*

Excavates horizontal burrow quite deep into steep wall of sand-pit, clayey river bank, etc., with nest chamber at end. Nest made of grass blades, fibres and feathers, contains 4—6 pure white eggs measuring 17× 13 mm.

36 Tree Pipit *(Anthus trivialis)*

All pipits build nest in depression in ground, well hidden by clump of grass, bush, heather or ferns. Tree Pipit's nest made of dry grass, large amount of moss and dead leaves, with softly lined cup. Clutch: 5 very variable, but mostly greyish white, brown-speckled eggs measuring 20×15 mm.

37 Red-backed Shrike *(Lanius collurio)*

Nest mostly built low in thicket, often in hedge. Made of stems, grass blades, roots and moss, lined with wool and fluff. Clutch: 4—6 whitish eggs with reddish brown spots (usually concentrated round blunt end). Size of eggs: 22×17 mm.

38 Lesser Whitethroat *(Sylvia curruca)*

Nest usually built low in dense shrubs (especially hedges), often in young conifers. Made of stems and grass blades, loosely put together. Clutch: 5 white eggs with sparse reddish brown spots. Size of eggs: 17×13 mm.

39 Dipper *(Cinclus cinclus)*

Nest large structure made of moss and plant parts in hole in bank or ground, under bridge or weir in mountain stream. Completely closed except for round entrance. Clutch: 4—6 pure white eggs measuring about 26×19 mm.

40 Hedge Sparrow — Dunnock *(Prunella modularis)*

Nest well hidden in thicket, close to ground. Made of few dry stems and large amount of green moss. Soft lining of cup usually includes red spore cases of moss. Clutch: 4—5 plain greenish blue eggs. Size of eggs: 19×14 mm.

41 Great Reed Warbler *(Acrocephalus arundinaceus)*

Nest built over water: Deep cup cleverly made of grass blades and rush fibres, carefully attached to reed blades. Clutch: 5—6 bluish white eggs with green to brown spots. Size of eggs: 23×16 mm.

42 Wood Warbler *(Phylloscopus sibilatrix)*

Spherical nest made of dry grass, with side entrance, built on ground. Clutch: 6—7 white, brown-spotted eggs. Size of eggs: 16×12 mm.

43 Icterine Warbler *(Hippolais icterina)*

Nest built in bush or tree, not too far from ground. Strongly made of plant and animal (spiders' webs) material, carefully smoothed and usually lined outside with thin slivers of birch bark. Clutch: 5 pink eggs with a few black spots. Size of eggs: 18.5×13.5 mm.

44 Chiffchaff *(Phylloscopus collybita)*

Closed nest with side entrance built in dense herbage or thicket on ground. Made of grass blades, dead leaves and moss, lined with feathers. Clutch: 6—7 white eggs with reddish brown spots. Size of eggs: 15×12 mm.

45 Willow Warbler *(Phylloscopus trochilus)*

Closed nest built (like that of all leaf warblers) well hidden on ground, usually in clump of grass or brushwood at edge of thicket. Oven-shaped, neatly made of grass blades, lined with leaves and feathers. Clutch: 6—7 whitish, russet-speckled eggs. Size of eggs: 15×12 mm.

46 Redstart *(Phoenicurus phoenicurus)*

Nest built in tree hollow, hole in wall or nest-box. Bowl-like structure made of grass blades, rootlets and leaves, lined with hair and feathers. Clutch: 5—7 greenish blue eggs measuring 19×14 mm.

47 Collared Flycatcher *(Ficedula albicollis)*

Nest loosely woven structure of all kinds of blades, rootlets and leaves, lined with hair and feathers, built in holes in various trees or in nest-boxes. Clutch: 5—7 bluish green eggs measuring 17×13 mm.

48 Black Redstart *(Phoenicurus ochruros)*

Nest shallow cup made of blades of grass, thin roots, plant fibres and dry leaves; cup is lined with hair and feathers. Stands in rock crevice, hole in wall, on roof-beam or even in room. Clutch: 5—6 white eggs measuring 19×14 mm.

49 Robin *(Erithacus rubecula)*

Nest built in hole in ground, hollow tree stump or side of ditch. Made of moss and dry leaves, with softly lined cup. Clutch: 5—6 yellowish eggs with yellow-brown speckles. Size of eggs: 20×15 mm.

50 Blackbird *(Turdus merula)*

Nest fairly large structure made of dry stalks, blades of grass and thin roots stuck together with mud. Deep cup. Built in bush, tree, building, rarely on ground as in picture (in sedge). Clutch: 4—6 bluish green eggs thickly covered with russet speckles. Size of eggs: 30×21 mm.

51 Fieldfare *(Turdus pilaris)*

Nest closely resembles Blackbird's, mostly built quite high up in tree. Deep cup often thickly lined with fine grass blades. Clutch: usually 5 eggs resembling Blackbird's, measuring 28.5×21 mm.

52 Great Tit *(Parus major)*

Nest made of blades of grass, roots, large amount of moss and lichen, built in any kind of tree hole or nest-box, sometimes in hole in wall, not too far from ground. Clutch: 8—10 white, red-spotted eggs. Size of eggs: 17×13 mm.

53 Nuthatch *(Sitta europaea)*

Nest loose litter of bark, dry leaves and plant fibres in hole in tree, nest-box or hole in wall. Entrance regularly made smaller with clay to fit bird's girth. Clutch: 6—8 white, russet-spotted eggs. Size of eggs: 19× 14.5 mm.

54 Yellowhammer *(Emberiza citrinella)*

Nest built mostly on ground, or in grass or thicket near ground. Made of blades of grass, roots and leaves, with hair-lined cup. Clutch: 4—5 whitish eggs with brown spots and scribbles. Size of eggs: 21.5× 16 mm.

55 Corn Bunting *(Emberiza calandra)*

Nest built on ground in field or meadow, concealed. Made of stems, grass leaves and rootlets loosely woven together, with softly lined cup. Clutch: 3—5 greyish yellow to reddish yellow eggs with brown spots and fine black scribbles and streaks. Size of eggs: 24× 17 mm.

56 Hawfinch *(Coccothraustes coccothraustes)*

Nest built mostly high up in deciduous tree. Made of brushwood, root fibres and moss, with fluff- and hair-lined cup. Clutch: 5 light blue eggs with a few black spots. Size of eggs: 24× 17.5 mm.

57 Chaffinch *(Fringilla coelebs)*

One of best constructed songbirds' nests. Fibres, blades of grass, moss and lichen firmly woven together, with deep, softly lined cup. Built mostly in tree, rarely in bush. Often made invisible by lichen outer lining. Clutch: 4—6 brownish eggs with characteristic dark 'burn' spots. Size of eggs: 19× 14.5 mm.

58 Siskin *(Carduelis spinus)*

Small, thick-walled, deep nest usually well hidden high up in conifer, at tips of twigs. Clutch: 4—6 whitish, russet-spotted eggs measuring 16× 12 mm.

59 Greenfinch *(Carduelis chloris)*

Bowl-shaped nest made of brushwood and blades, with softly lined cup. Usually not very high up, in hedge, bush or young conifer. Clutch: 5—6 whitish eggs with dark brown spots and speckles. Size of eggs: 20× 14.5 mm.

60 Bullfinch *(Pyrrhula pyrrhula)*

Shallow nest made of brushwood, roots and lichen. Built not too high up in conifer, or in thicket at edge of wood or clearing. Complete clutch comprises 4—5 blue eggs with a few blackish brown spots. Size of eggs: 20.5× 15 mm.

61 Tree Sparrow *(Passer montanus)*

Typical nest usually untidy mass of straw, blades of grass, roots and feathers with warmly lined, spherical cup, in tree hole, nest-box or hole in wall. Clutch: 5—6 whitish, thickly brown-spotted eggs measuring 19 × 14 mm.

62 Magpie - Black-billed Magpie *(Pica pica)*

Characteristic large nest, usually with arched roof, built in tree, occasionally in thicket. Made of twigs and brushwood, with blades of grass and hair-lined well. Clutch: 6—7 greenish eggs thickly covered with dark-brown spots and speckles. Size of eggs: 33 × 23 mm.

63 Golden Oriole *(Oriolus oriolus)*

Beautifully constructed nest made of grass blades, phloem fibres, moss, wool and feathers woven together, with deep cup. Built in fork of branch, usually high up near outer margin of tree. Clutch: 4 white eggs with a few brownish black spots. Size of eggs: 30 × 21 mm.

64 Carrion Crow *(Corvus corone)*

Nest large dish built high up in top of tree. Made of twigs, brushwood and tufts of grass, often reinforced with soil, with moss- and hair-lined well. Clutch: 5 bluish green eggs covered with brownish black spots. Size of eggs: 41 × 29 mm.

BIRDS NESTING IN MARGINAL REGIONS AND A FEW ACCIDENTAL BIRDS

Apart from the 336 species illustrated in the book, there are a number of birds, nesting mainly in the Mediterranean countries and eastern Europe, for which we were unable to find room. Since they belong to the regular bird population of Europe and/or North Africa and Asia Minor, we felt we ought to give at least a brief description of them. In addition, certain accidental birds which stray from their normal route while migrating from their home in Asia or North America, or from their winter area, can also be encountered, in exceptional cases, in Europe or in European waters. Of these we have mentioned only the commonest.

Family: Shearwaters - Procellariidae

Manx Shearwater *(Puffinus puffinus)*. Slightly smaller than Black-headed Gull. Blackish brown, sides of neck grey, underside white. Black beak. Nests in Atlantic and Mediterranean region.

Sooty Shearwater *(Puffinus griseus)*. About size of Common Gull. Blackish grey, with dark beak. Nests in southern hemisphere, roams north Atlantic.

Fulmar *(Fulmarus glacialis)*. Size of large Common Gull. Grey back, white or grey head and tail. Stocky body, large head, yellow beak. Nests in colonies in north Atlantic and Arctic Ocean.

Family: Storm Petrels - Hydrobatidae

Storm Petrel *(Hydrobates pelagicus)*. Smaller than Swallow. Sooty black, with gleaming white rump, black legs. Flutters on surface of water. Nests in Atlantic and Mediterranean region, frequently driven inland.

Leach's Petrel *(Oceanodroma leucorrhoa)*. Size of large Swallow. Brownish black, white rump with grey centre, notched tail. Flies over water in 'hops'. Atlantic Ocean.

Madeiran Petrel *(Oceanodroma castro)*. Like preceding species, but pure white rump. Atlantic Ocean.

Family: Gannets - Sulidae

Gannet *(Sula bassana)*. Size of goose. Large head, thick neck, thick wedge-shaped beak. White, with black primaries. Juvenile brown or spotted. Nests in north Atlantic region in large colonies.

Family: Cormorants - Phalacrocoracidae

Shag *(Phalacrocorax aristotelis)*. Resembles Cormorant, but smaller, with small crest on crown. Sea-bird, nests in colonies on cliffs in north Atlantic and Mediterranean.

Pygmy Cormorant *(Phalacrocorax pygmeus)*. About half size of Cormorant. Dark, greenish black plumage and russet head. Nests beside inland water from southern Europe to Kazakhstan and in North Africa.

Family: Pelicans - Pelecanidae

White Pelican *(Pelecanus onocrotalus)*. Size of swan. Long, powerful beak. Retracts head when flying. White; juvenile brown-spotted. Underside of wing feathers black. Legs pink. Danube delta to inner Asia.

Dalmatian Pelican *(Pelecanus crispus)*. Closely resembles White Pelican, but underside of wings mainly white (only tips of feathers dark brown). Legs lead grey. Distribution as for White Pelican.

Family: Herons - Ardeidae

Cattle Egret *(Ardeola ibis)*. Resembles Squacco Heron, but whiter. Beak and legs red in nuptial plumage, otherwise yellow or dark. Found mostly among grazing cattle. South of Spain, Africa, Mesopotamia.

Family: Flamingos - Phoenicopteridae

Greater Flamingo *(Phoenicopterus ruber)*. Slim body, very long neck, long legs. White, with scarlet and black wings. Flies with neck and legs extended. Massive, curved beak. South of France, Spain, North Africa, inner Asia.

Family: Ducks - Anatidae

American Widgeon (Baldpate) *(Anas americana)*. Like Widgeon. White crown, dark green band along sides of head. North America, frequent accidental bird in Europe.

Marbled Teal *(Anas angustirostris)*. Larger than Teal. Short crest, dark spot on sides of head, light brown, speckled body, white tail. Mediterranean region and Near East.

Mandarin *(Aix galericulata)*. Larger than Teal. Gay, white-sided crest, cinnamon side 'whiskers', cinnamon primaries raised like sails. Eastern Asia. Introduced into Europe as ornamental bird, often nests in open, in tree hole.

King Eider *(Somateria spectabilis)*. Smaller than Eider. Breast and front of body white, remainder black. Large, grey head with orange patch on forehead. Nests on arctic coasts.

Harlequin *(Histrionicus histrionicus)*. Larger than Teal. Head, breast and end of body black, vividly marked with white stripes and spots. Back bluish grey, sides rust brown. Iceland, Greenland, Arctic region; often occurs in western Europe.

Barrow's Goldeneye *(Bucephala islandica)*. Resembles Goldeneye, but has crescent-shaped spot on cheeks. Nests in Iceland and North America; recurring accidental bird in western Europe.

White-headed Duck *(Oxyura leucocephala)*. About size of Pochard. Erect, pointed tail. Brown, with white head and blue bill. Southern European inland waters; accidental bird in central and western Europe.

Snow Goose *(Anser caerulescens)*. Size of goose, snow white with black primaries. Reddish bill. Do not confuse with albino Greylag and Bean Geese (without black wing tips)! North Siberian tundra; in winter on north Atlantic coast.

Canada Goose *(Branta canadensis)*. Larger than Greylag Goose. Black head and neck, white cheeks, brown back. Black bill and legs. North America; now settled in northern Europe.

Red-breasted Goose *(Branta ruficollis)*. Smaller than domestic goose. Body black and white, front of neck and breast chestnut. Short bill and legs black. North Asian tundra; often strays into Europe.

Ruddy Shelduck *(Casarca ferruginea)*. Smaller than domestic goose. Reddish brown with whitish brown head and white wing coverts. North-west Africa, south of Spain. Black Sea region to eastern Asia.

Family: Vultures, Eagles, Buzzards, etc. — Accipitridae

Egyptian Vulture *(Neophron percnopterus)*. Larger than Common Buzzard. Head and throat unfeathered, plumage dingy white, black-tipped wings. White, triangular tail. Mediterranean region, Asia, Africa.

Bearded Vulture (Lammergeier) *(Gypaëtus barbatus)*. About size of White-tailed Eagle. Light head and neck, rusty yellow underside (in juvenile dark). Narrow, pointed wings, long, triangular tail. Rocky mountains in southern Europe and Asia.

Steppe Eagle *(Aquila rapax)*. Somewhat larger than similar Lesser Spotted Eagle. Entirely brown; short, rounded tail with indistinct grey cross-bands. Steppes betweeen Black Sea and inner Asia, also Africa.

Bonelli's Eagle *(Hieraaëtus fasciatus)*. Smallish eagle with slim body and long tail, resembles Goshawk. White underside, dark wings, cross-banded tail. Mediterranean region, southern Asia, Africa.

Levant Sparrow Hawk *(Accipter brevipes)*. Closely resembles Sparrow Hawk. Black-tipped wings with white under surface. Grey cheeks. Balkan Peninsula, southern Asia, Africa.

Black-winged Kite *(Elanus caeruleus)*. Size of Kestrel. Light grey, white tail, white shoulders. South of Portugal, tropical grasslands.

Family: Falcons - Falconidae

Eleonora's Falcon *(Falco eleonorae)*. Slightly larger than similar Hobby. Lacks russet 'trousers'. Sometimes entirely brownish black, without visible markings. Lives in colonies on rocky islands in Mediterranean.

Lanner Falcon *(Falco biarmicus)*. Resembles Peregrine Falcon, but lighter, with sandy crown, brown back and spotted underside. Mediterranean region and Africa.

Gyr Falcon *(Falco rusticolus)*. Much larger and stronger than Peregrine Falcon. Light grey, sometimes white, with no moustachial stripe. Rocky coastal regions of northern Europe, Greenland and North America.

Lesser Kestrel *(Falco naumanni)*. Distinguished from similar Kestrel by unspotted back, bluer head and tail and gregarious habits. Mediterranean region to inner Asia.

Family: Game Birds - Phasianidae

Barbary Partridge *(Alectoris barbara)*. Resembles Red-legged Partridge, but has grey chin, grey-sided head and brown, white-speckled collar. North Africa, Sardinia.

Black Francolin *(Francolinus francolinus)*. About size of Partridge. Back dark brown, head and sides of neck black, with white patch in middle, underside black, with white spots. Asia Minor to India.

Family: Rails - Rallidae

Purple Gallinule *(Porphyrio porphyrio)*. Size of large Coot. Glossy deep blue, with large, red beak and red frontal patch. Long, red legs. Mediterranean region, Africa, southern Asia, Australia.

Family: Cranes - Gruidae

Demoiselle Crane *(Anthropoides virgo)*. Smaller than Crane. Grey, front of neck and breast black, long breast feathers. Tufts of white feathers on sides of head. Steppes from Danube delta to eastern Asia.

Family: Bustards - Otididae

Houbara Bustard *(Chlamydotis undulata)*. Resembles hen turkey. Short, black and white crest, tufts of black feathers down sides of neck. Africa and southern Asia; frequent accidental bird in Europe.

Family: Plovers and Lapwings - Charadriidae

Spur-winged Plover *(Vanellus spinosus)*. Smaller than Lapwing. Black and white, fawn back, black head. Africa, Near East, Greece.

Sociable Plover *(Vanellus gregarius)*. Size of Lapwing. Black crown without crest, broad, white frontal and eye stripes. Lower breast brown, rump and underside of tail white. Asia steppes; frequent accidental bird in Europe.

Family: Snipes and Sandpipers - Scolopacidae

Slender-billed Curlew *(Numenius tenuirostris)*. Distinguished from very similar Curlew by round, blackish brown spots on sides. Rump pure white. Western Siberia, winters in eastern Mediterranean region; accidental bird in western Europe.

Lesser Yellow-legs *(Tringa flavipes)*. Almost as large as Redshank. Fairly long beak, yellow legs. Rump white, no white on wings. North America; frequent accidental bird in western Europe.

Terek Sandpiper *(Tringa cinereus)*. Smaller than Redshank, with thin, upcurved beak. Yellow legs, white wing bars, light rump. Northern Europe, northern Siberia; often seen in rest of Europe.

Purple Sandpiper *(Calidris maritima)*. Larger than Dunlin. Slate grey, with spotted back. Base of beak and short legs yellow. Iceland and northern Scandinavia, in winter on west European coasts; inland as accidental bird.

Broad-billed Sandpiper *(Limicola falcinellus)*. Smaller than Dunlin. Short legs. White eye stripe forks behind eyes. Back similar to Snipe's. North Scandinavia; migrates across Europe.

Family: Pratincoles - Glareolidae

Cream-coloured Courser *(Cursorius cursor)*. Size of thrush. Long legs, moderately long, gently curved beak. Light sandy, with white and black eye stripes. Legs yellow. African and west Asian deserts; frequent accidental bird in whole of Europe.

Family: Gulls - Laridae

Iceland Gull *(Larus glaucoides)*. Size of Herring Gull, in colouring like small Glaucous Gull. Arctic North America and Greenland, infrequent winter bird in North Sea; accidental bird on west European coasts.

Audouin's Gull *(Larus audouini)*. Smaller than Herring Gull. Bright red beak with black ring. Legs dark green. Black wing tips, not sharply defined. A few Mediterranean islands (Corsica, Sardinia, Cyprus).

Slender-billed Gull *(Larus genei)*. Similar wing markings to Black-headed Gull, but with white head and neck, and longer, coral red beak. Mediterranean to inner Asia.

Great Black-headed Gull *(Larus ichthyaëtus)*. Size of Great Black-backed Gull. Black head, yellow, black-ringed beak, greenish yellow legs. Nests between Black Sea and steppe lakes of inner Asia, winters partly in eastern Mediterranean.

Sabine's Gull *(Xema sabini)*. Smaller than Black-headed Gull. Forked tail, greyish brown head (in winter white), black-tipped wings with white posterior border. Arctic region; accidental bird on west European coasts.

Ross's Gull *(Rhodostethia rosea)*. Smaller than Black-headed Gull. Triangular tail with pointed tip. Black collar, pink underside. Arctic region; accidental bird in western and northern Europe.

Family: Sandgrouse - Pteroclidae

Pin-tailed Sandgrouse *(Pterocles alchata)*. About size of pigeon. Short beak and legs, dagger-like tail. Sandy back, with spots and stripes. Ochre brown breast bounded by narrow black bands. Belly white. Dry parts of Spain, North Africa and Asia Minor.

Black-bellied Sandgrouse *(Pterocles orientalis)*. Larger than Pin-tailed Sandgrouse, distinguished by black belly and blackish under surface of wings. Spain, Portugal, North Africa, Asia Minor.

Family: Cuckoos - Cuculidae

Great Spotted Cuckoo *(Clamator glandarius)*. Size of large Cuckoo. Grey, white-spotted back, light grey crest, white underside, long, gradated tail. Southern Europe, Africa, Asia Minor.

Family: Owls - Strigidae

Great Grey Owl *(Strix nebulosa)*. Slightly smaller than Eagle Owl. Grey, spotted; broad stripes down underside, concentric rings on face. Yellow eyes. Scandinavia, northern Asia, North America.

Family: Swifts - Apodidae

Pallid Swift *(Apus pallidus)*. Like Swift in flight. Brown, with white throat. Belly dark brown. Mediterranean region.

Family: Larks - Alaudidae

Calandra Lark *(Melanocorypha calandra)* Larger and more ungainly than similar Skylark. Thick beak, large white spot on either side of neck. Sings while flying. Nests in steppes of Mediterranean countries and inner Asia; accidental bird in central and western Europe.

Black Lark *(Melanocorypha yeltoniensis)*. Large lark. ♂ black, ♀ sandy. Large yellow beak. Steppes between river Don and Mongolia; rare accidental bird in central and western Europe.

White-winged Lark *(Melanocorypha leucoptera)*. Large lark. Back light brown, striped. Russet crown, white underside. Broad white specula. Nests between river Don and Mongolia; accidental bird in Europe.

Lesser Short-toed Lark *(Calandrella rufescens)*. Distinguished from very similar Short-toed Lark by finely striped breast. South of Spain, North Africa, Asia.

Thekla Lark *(Galerida theklae)*. Very hard to distinguish from Crested Lark. Stripes on breast darker and more pronounced, beak rather longer. Iberian Peninsula, Balearic Islands, south of France.

Family: Titmice - Paridae

Azure Tit *(Parus cyanus)*. Larger than similar Blue Tit. White head and underside, greyish blue back, dark blue eye stripe. U.S.S.R. and North China.

Siberian Tit *(Parus cinctus)*. Resembles large Marsh Tit. Brownish black cap, large chin spot, brown back. Northern Europe, northern Asia.

Sombre Tit *(Parus lugubris)*. Size of Great Tit, closely resembles Marsh Tit. Large brown chin spot, black cap, greyish brown back, strong beak. Balkan Peninsula, Asia Minor.

Family: Nuthatches - Sittidae

Corsican Nuthatch *(Sitta whiteheadi)*. Small nuthatch with black crown and eye stripe and white band above eyes. Only on Corsica.

Rock Nuthatch *(Sitta neumayer)*. Resembles Nuthatch, but lighter coloured, with white underside. Balkan Peninsula and Near East.

Family: Thrushes - Turdidae

White's Thrush *(Zoothera dauma)*. Resembles large Mistle Thrush, but more golden brown; with wide, scale-like spots. Broad white band on dark under surface of wings. Siberia, accidental bird in Europe.

Siberian Thrush *(Turdus sibiricus)*. Size of Song Thrush. Slate black, with vivid white eye stripes. Siberia; accidental bird in Europe.

Eye-browed Thrush *(Turdus obscurus)*. Smaller than Song Thrush. Greyish brown, with rust brown sides. White eye stripe, white belly. Siberia; accidental bird in Europe.

Naumann's Thrush *(Turdus naumanni)*. Average thrush size. Throat, breast, rump and tail russet, wide russet border on primaries. White belly. More southern variety, Dusky Thrush *(T. n. eunomus)*, has black-spotted breast, black crown and tail and rust brown wings. Siberia; accidental bird in Europe.

Black-throated Thrush *(Turdus ruficollis)*. Average thrush size. Brownish grey with russet (variety *ruficollis*) or black throat and breast (var. *atrigularis*). Belly white. Siberia; accidental bird in Europe.

American Robin *(Turdus migratorius)*. Size of Blackbird. Black back, bright russet breast, white, black-striped chin, yellow beak. North America; accidental bird in Europe.

Isabelline Wheatear *(Oenanthe isabellina)*. Larger than Wheatear, longer beak. Sandy, without black head markings. Tail and wing tips black, rump and belly white. Asia Minor to North China; accidental bird in Europe.

Pied Wheatear *(Oenanthe pleschanka)*. Same size as Wheatear. White crown, rump and belly. Back, throat, wings and tail black. West coast of Black Sea to Mongolia; accidental bird in Europe.

Black Wheatear *(Oenanthe leucura)*. Larger than Wheatear. All black, except for white rump and outer tail feathers. Iberian Peninsula, North Africa.

Red-flanked Bluetail *(Tarsiger cyanurus)*. Size of Robin. ♂: deep blue back, white underside, orange red sides. ♀: olive brown, rump and tail blue. From Finland, across northern Russia, to Japan; accidental bird in northern and western Europe.

Family: Warblers - Sylviidae

Olive-tree Warbler *(Hippolais olivetorum)*. Larger than Icterine Warbler, with longer, thicker beak. Grey, with white underside and white-edged wing feathers. Greece, Asia Minor.

Rüppell's Warbler *(Sylvia rüppelli)*. Large warbler. Black head and throat, white moustachial stripe, dark grey back, white underside. Red legs. Greece, Asia Minor.

Spectacled Warbler *(Sylvia conspicillata)*. Quite small, resembles Whitethroat, but darker head and redder breast. Mediterranean region.

Marmora's Warbler *(Sylvia sarda)*. Small, slate black warbler with erect tail. Red legs and orbital ring. West Mediterranean region.

Rufous Warbler (Rufous Bush Chat) *(Cercotrichas galactotes)*. Size of large Sparrow. Rust brown, underside white. Fanned, cocked tail with white and black border. Tuneful song sung from perch in open. Southern Europe, Near East, Africa; accidental bird in western Europe.

Yellow-browed Warbler *(Phylloscopus inornatus)*. Distinguished from similar Willow Warbler by double wing bar and more vivid wing markings. Northern and inner Asia; accidental bird in northern and central Europe.

Pallas's Warbler *(Phylloscopus proregulus)*. Size of Goldcrest. Green back, yellow rump, double wing bands, vivid head markings (yellow and black eye stripes). East Siberia; accidental bird in northern and western Europe.

Family: Wagtails - Motacillidae

Richard's Pipit *(Anthus novaeseelandiae)*. Size of White Wagtail. Long legs with long-clawed hind toe. Dark; back and breast striped. Asia, Africa, Australia. Almost regular accidental bird in western and northern Europe.

Family: Shrikes - Laniidae

Masked Shrike *(Lanius nubicus)*. Size of Red-backed Shrike, but black, with white patch on wings, white forehead and white throat. Rust-coloured sides. Greece and Asia Minor.

Family: Finches - Fringillidae

Trumpeter Bullfinch *(Rhodopechys githaginea)*. Slightly smaller than Sparrow. Short, thick, coral beak. Grey plumage, with crimson head and pink underside. Iberian Peninsula, North Africa, Greece to Inner Asia.

Family: Buntings - Emberizidae

Pine-bunting *(Emberiza leucocephala)*. Size of Yellowhammer. Crown and belly white, throat, breast, wide eye stripe and rump chestnut. East Russia to west China; accidental bird in Europe.

Red-headed Bunting *(Emberiza bruniceps)*. Size of Black-headed Bunting, but chestnut (not black) head. Transcaspian region; frequent accidental bird in western and central Europe, in recent years some birds escaped from captivity.

Yellow-breasted Bunting *(Emberiza aureola)*. Size of Sparrow. Brown back, black head, yellow underside with brown band on breast. Finland to eastern Asia; accidental bird in central and western Europe.

Cretzschmar's Bunting *(Emberiza caesia)*. Closely resembles Ortolan Bunting, but with cinnamon (not yellow) throat and bluish grey head and breast-band. Balkan Peninsula and Asia Minor; accidental bird in western Europe.

Rustic Bunting *(Emberiza rustica)*. Size of Sparrow. Rust brown back and wide breast-band, white throat and belly, black head. Scandinavia and northern Asia; accidental bird in central and western Europe.

Little Bunting *(Emberiza pusilla)*. Smaller than Sparrow. Russet, black-striped crown and cheeks. Finland to eastern Siberia; accidental bird in western and central Europe.

BIBLIOGRAPHY

Alexander, W. B.: Birds of the Ocean. London 1954
Austin, O. L.: Birds of the World. London 1961

Bauer, K. M. and Glutz N.: Handbuch der Vögel Mitteleuropas, 5 vols. Frankfurt 1966—1973
Beme, P. L. and Kusnetzow, A. A.: Forest and Hill Birds of the USSR. Moscow 1966
Berndt, R. and Meise, W.: Naturgeschichte der Vögel, 3 vols. Stuttgart 1959—1966
Bruun, B. and Singer, A.: The Hamlyn Guide to Birds of Britain and Europe. London 1970
Bruun, B. and Singer, A.: British and European Birds in Colour. London 1969

Dementiev, G. P. and Gladkov, N. A.: Birds of the USSR, 6 vols. Moscow 1951—1954

Gooders, J.: Where to Watch Birds in Britain and Europe. London 1970

Heinroth, O. and M.: Die Vögel Mitteleuropas, 4 vols. Frankfurt 1966—1967
Hollom, P. A. D.: The Popular Handbook of Rarer British Birds, 4th edition. London 1970

Makatsch, W.: Wir bestimmen die Vögel Europas. Melsungen 1966

Nørevang, A. and Meyer, T. J.: Wir beobachten Vögel. Jena 1965

Peterson, R., Mountfort, G. and Hollom, P. A. D.: A Field Guide to the Birds of Britain and Europe. London 1954

Robbins, C. S., Bruun, B., Zim, H. S. and Singer, A.: A Guide to Field Identification: Birds of North America. New York 1966

Stresemann, E. and Portenko, L. A.: Atlas der Verbreitung palaearktischer Vögel. Berlin 1960 to 1974

Vaurie, Ch.: The Birds of the Palaearctic Fauna, 2 vols. London 1959—1965
Voous, K. H.: Atlas of European Birds. Edinburgh 1960

Witherby, H. F. and coll.: The Handbook of British Birds, 5 vols., London 1949

INDEX OF COMMON BIRDS' NAMES

Alpine Accentor 172
Alpine Chough 244
Alpine Swift 146
American Oystercatcher 90, 272
American Robin 327
American Widgeon 322
Aquatic Warbler 178
Arctic Loon 24
Arctic Skua 114, 284, 315
Arctic Tern 126, 292
Audouin's Gull 325
Avocet 110, 272
Azure Tit 326
Azure-winged Magpie 242

Baillon's Crake 88
Baldpate 322
Bank Swallow 160, 316
Barbary Partridge 324
Barnacle Goose 54, 258
Barn Owl 136
Barn Swallow 158, 316
Barred Warbler 188
Barrow's Goldeneye 322
Bar-tailed Godwit 100
Bean Goose 54, 256
Bearded Titmouse 208
Bearded Vulture 323
Bee-eater 148
Bewick's Swan 56, 258
Bittern 30
Black-bellied Plover 94
Black-bellied Sandgrouse 325
Black-billed Magpie 242, 320
Blackbird 208, 318
Blackcap 184
Black-crowned Night Heron 34
Black-eared Wheatear 202
Black Francolin 324
Black Grouse 78
Black Guillemot 132, 315
Black-headed Bunting 222
Black-headed Gull 120, 288, 315
Black Kite 68, 266
Black Lark 326
Black-legged Kittiwake 122, 288
Black-necked Grebe 28
Black-necked Stilt 110, 272
Black Redstart 198, 318
Black Stork 36
Black-tailed Godwit 100, 315
Black Tern 124, 294
Black-throated Diver 24
Black-throated Thrush 327
Black Vulture 58, 260
Black Wheatear 327
Black-winged Kite 323
Black-winged Stilt 110, 272
Black Woodpecker 150
Blue-headed Wagtail 162, 316
Blue Rock Thrush 204
Bluethroat 200
Blue Tit 212
Bohemian Waxwing 168
Bonelli's Eagle 323
Bonelli's Warbler 190
Booted Eagle 62, 262
Boreal Owl 142
Brambling 224
Brant Goose 54, 258
Brent Goose 54, 258
Broad-billed Sandpiper 325
Brown Creeper 216
Bullfinch 234, 319
Buzzard 64, 264

Calandra Lark 326
Canada Goose 323
Capercaillie 78
Carrion Crow 246, 320
Caspian Tern 122, 290
Cattle Egret 322
Cetti's Warbler 174
Chaffinch 224, 319
Chiffchaff 188, 317
Chough 244
Chukar 80
Cirl Bunting 218

Citril Finch 228
Coal Tit 210
Collared Flycatcher 194, 318
Collared Turtle Dove 134
Common Egret 32
Common Eider 46, 252
Common Gallinule 86, 314
Common Goldeneye 42, 252
Common Gull 120, 286
Common Loon 24
Common Merganser 50, 254
Common Murre 130
Common Puffin 130
Common Raven 246
Common Redpoll 226
Common Sandpiper 104, 278
Common Scoter 48, 254
Common Snipe 96, 314
Common Tern 126, 292, 315
Coot 84, 314
Cormorant 28
Corn Bunting 218, 319
Corn Crake 86
Corsican Nuthatch 327
Crag Martin 160
Crane 82
Cream-coloured Courser 325
Crested Lark 156
Crested Tit 210
Cretzschmar's Bunting 329
Crossbill 232
Cuckoo 144
Curlew 98, 315
Curlew Sandpiper 108, 280

Dalmatian Pelican 322
Dartford Warbler 186
Demoiselle Crane 324
Dipper 170, 317
Dotterel 94, 276
Dunlin 108, 280
Dunnock 172, 317

Eagle Owl 136
Eared Grebe 28
Egyptian Vulture 323
Eider 46, 252
Eleonora's Falcon 323
European Goldfinch 226
European Widgeon 42, 250
Eye-browed Thrush 327
Fan-tailed Warbler 174
Ferruginous Duck 46, 250
Fieldfare 206, 318
Firecrest 192
Fulmar 321

Gadwall 38, 248
Gannet 321
Garden Warbler 182
Garganey 40, 248
Glaucous Gull 116, 286
Glossy Ibis 34
Goldcrest 192
Golden Eagle 60, 260
Goldeneye 42, 252
Golden Oriole 240, 320
Golden Plover 94, 274
Goldfinch 226
Goosander 50, 254
Goshawk 66, 266
Grasshopper Warbler 176
Gray Partridge 80, 313
Great Black-backed Gull 118, 286
Great Black-headed Gull 325
Great Bustard 84
Great Cormorant 28
Great Crested Grebe 26, 313
Greater Flamingo 322
Greater Scaup 44, 252
Great Grey Owl 326
Great Grey Shrike 168
Great Northern Diver 24
Great Reed Warbler 178, 317
Great Skua 116, 284
Great Snipe 96
Great Spotted Cuckoo 326
Great Spotted Woodpecker 152
Great Tit 214, 318
Great White Egret 32
Greenfinch 224, 319
Green Sandpiper 104, 278
Greenshank 102, 278
Green-winged Teal 40, 248
Green Woodpecker 150
Greenish Warbler 192
Grey-headed Woodpecker 150
Grey Heron 30
Greylag Goose 52, 256
Grey Phalarope 112, 282
Grey Plover 94, 274

Grey Wagtail 162
Griffon Vulture 58, 260
Guillemot 130
Gull-billed Tern 126, 290
Gyr Falcon 324

Harlequin 322
Hawk Owl 142
Hawfinch 222, 319
Hazel Hen 78
Hedge Sparrow 172, 317
Hen Harrier 70, 268
Herring Gull 118, 286
Hobby 74, 270
Honey Buzzard 66, 264
Hooded Crow 246
Hoopoe 144
Horned Grebe 28
Horned Lark 158
Houbara Bustard 324
House Martin 160
House Sparrow 236

Iceland Gull 325
Icterine Warbler 180, 317
Imperial Eagle 60, 260
Isabelline Wheatear 327

Jackdaw 244
Jack Snipe 96
Jay 240

Kentish Plover 92, 274
Kestrel 74, 270, 313
King Eider 322
Kingfisher 148
Kittiwake 122, 288
Knot 106, 280

Lammergeier 323
Lanner Falcon 323
Lapland Bunting 222
Lapland Longspur 222
Lapwing 90, 272, 314
Leach's Petrel 321
Least Tern 128, 290
Lesser Black-backed Gull 118, 286
Lesser Grey Shrike 168
Lesser Kestrel 324
Lesser Short-toed Lark 326
Lesser Spotted Eagle 62, 262
Lesser Spotted Woodpecker 154
Lesser White-fronted Goose 52, 256
Lesser Whitethroat 186, 317
Lesser Yellow-legs 324
Levant Sparrow Hawk 323
Linnet 228
Little Bittern 34, 313
Little Bunting 329
Little Bustard 84
Little Crake 88, 314
Little Egret 32
Little Grebe 26
Little Gull 122, 288
Little Owl 138
Little Ringed Plover 92, 274
Little Stint 106, 280
Little Tern 128, 290
Long-eared Owl 138, 316
Long-legged Buzzard 64, 264
Long-tailed Duck 46, 252
Long-tailed Jaeger 116, 284
Long-tailed Skua 116, 284
Long-tailed Tit 208

Madeiran Petrel 321
Magpie 242, 320
Mallard 38, 248, 313
Mandarin 322
Manx Shearwater 321
Marbled Teal 322
Marmora's Warbler 328
Marsh Harrier 68, 268, 313
Marsh Hawk 70, 268
Marsh Sandpiper 102, 276
Marsh Tit 212
Marsh Warbler 180
Masked Shrike 328
Meadow Pipit 166
Mediterranean Gull 120, 288
Melodious Warbler 182
Merlin 74, 270
Mew Gull 120, 286
Middle Spotted Woodpecker 152
Mistle Thrush 204
Montagu's Harrier 70, 268
Moorhen 86, 314
Moustached Warbler 174
Mute Swan 56, 258, 313

Naumann's Thrush 327
Night Heron 34

Nightingale 200
Nightjar 144
Northern Phalarope 112, 282
Northern Shrike 168
Northern Three-toed Woodpecker 154
Nutcracker 242
Nuthatch 214, 319

Oldsquaw 46, 252
Olivaceous Warbler 182
Olive-tree Warbler 327
Orphean Warbler 184
Ortolan Bunting 220
Osprey 72, 262
Oystercatcher 90, 272

Pallas's Warbler 328
Pallid Harrier 70, 268
Pallid Swift 326
Parasitic Jaeger 114, 284, 315
Parrot Crossbill 232
Partridge 80, 313
Penduline Tit 210
Peregrine Falcon 72, 270
Pheasant 82, 314
Pied Flycatcher 194
Pied Wheatear 327
Pigeon Hawk 74, 270
Pine-bunting 328
Pine Grosbeak 230
Pintail 38, 248, 313
Pin-tailed Sandgrouse 325
Pochard 44, 250
Pomarine Jaeger 114, 284
Pomarine Skua 114, 284
Pratincole 112
Ptarmigan 76
Puffin 130
Purple Gallinule 324
Purple Heron 30
Purple Sandpiper 325
Pygmy Cormorant 322
Pygmy Owl 140

Quail 82

Raven 246
Razorbill 130, 315
Red-backed Shrike 170, 317
Red-breasted Flycatcher 196
Red-breasted Goose 323
Red-breasted Merganser 50, 254
Red-crested Pochard 42, 250
Red Crossbill 232
Red-flanked Bluetail 327
Red-footed Falcon 76, 270
Red-headed Bunting 328
Red Kite 68, 266
Red-legged Partridge 80
Red-necked Grebe 26
Red-necked Phalarope 112, 282
Red Phalarope 112, 282
Redpoll 226
Redstart 198, 318
Red-rumped Swallow 162
Redshank 102, 276, 315
Red-throated Diver 24
Red-throated Loon 24
Red-throated Pipit 166
Redwing 206
Reed Bunting 220
Reed Warbler 180
Richard's Pipit 328
Ringed Plover 92, 274, 314
Ring-necked Pheasant 82, 314
Ring Ouzel 206
River Warbler 176
Robin 198, 318
Rock Nuthatch 327
Rock Ptarmigan 76
Rock Bunting 220
Rock Dove 132
Rock Partridge 80
Rock Sparrow 234
Rock Thrush 202
Roller 146
Rook 246
Roseate Tern 128, 292
Rose-coloured Starling 238
Ross's Gull 325
Rough-legged Buzzard 64, 264
Rough-legged Hawk 64, 264
Ruddy Shelduck 323
Ruddy Turnstone 90, 314
Ruff 110, 282, 315
Rufous Bush Chat 328
Rufous Warbler 328
Rüppell's Warbler 328
Rustic Bunting 329

Sabine's Gull 325
Saker Falcon 72, 270

Sanderling 108, 282
Sand Martin 160, 316
Sandwich Tern 128, 290
Sardinian Warbler 184
Savi's Warbler 176
Scarlet Grosbeak 230
Scaup 44, 252
Scops Owl 140
Sedge Warbler 178
Serin 230
Shag 322
Shelduck 48, 256
Shore Lark 158
Short-eared Owl 138
Short-toed Eagle 62, 262
Short-toed Lark 158
Short-toed Tree Creeper 216
Shoveler 40, 250
Siberian Jay 240
Siberian Thrush 327
Siberian Tit 326
Siskin 226, 319
Skua 116, 284
Skylark 156, 316
Slavonian Grebe 28
Slender-billed Curlew 324
Slender-billed Gull 325
Smew 50, 254
Snipe 96, 314
Snow Bunting 216
Snow Finch 234
Snow Goose 323
Snowy Owl 136
Snowy Plover 92
Sociable Plover 324
Sombre Tit 326
Song Thrush 204
Sooty Shearwater 321
Spanish Sparrow 236
Sparrow Hawk 66, 266
Spectacled Warbler 328
Spoonbill 36
Spotless Starling 238
Spotted Eagle 60, 262
Spotted Sandpiper 104, 278
Spotted Crake 88
Spotted Flycatcher 194
Spotted Redshank 100, 276
Spur-winged Plover 324
Squacco Heron 32
Starling 238
Steppe Eagle 323
Stock Dove 132
Stonechat 196
Stone Curlew 114, 282, 315
Storm Petrel 321
Subalpine Warbler 188
Swallow 158, 316
Swift 146
Syrian Woodpecker 152

Tawny Owl 140
Tawny Pipit 164
Teal 40, 248
Temminck's Stint 106, 280
Tengmalm's Owl 142
Terek Sandpiper 325
Thekla Lark 326
Three-toed Woodpecker 154
Thrush Nightingale 200
Tree Creeper 216
Tree Pipit 166, 316
Tree Sparrow 236, 320
Trumpeter Bullfinch 328
Tufted Duck 44, 252
Turnstone 90, 272, 314
Turtle Dove 134
Twite 228
Two-barred Crossbill 232

Ural Owl 142

Velvet Scoter 48, 254

Wall Creeper 214
Water Pipit 164
Water Rail 86
Waxwing 168
Wheatear 202
Whimbrel 98
Whinchat 196
Whiskered Tern 124, 294
Whistling Swan 56, 258
White-backed Woodpecker 154
White-fronted Goose 52, 256
White-headed Duck 323
White Pelican 322
White's Thrush 327
White Stork 36
White-tailed Eagle 58, 260
Whitethroat 186
White Wagtail 164, 316

White-winged Black Tern 124, 294
White-winged Crossbill 232
White-winged Lark 326
White-winged Scoter 48, 254
Whooper Swan 56, 258
Wigeon 42, 250
Willow Grouse 76
Willow Ptarmigan 76
Willow Tit 212
Willow Warbler 190, 318
Winter Wren 172
Woodchat Shrike 170
Woodcock 98, 314
Wood Lark 156
Wood Pigeon 134, 316
Wood Sandpiper 104, 278
Wood Warbler 190, 317
Wren 172
Wryneck 148, 316

Yellow-breasted Bunting 329
Yellow-browed Warbler 328
Yellowhammer 218, 319
Yellow Wagtail 162, 316

INDEX OF SCIENTIFIC BIRDS' NAMES

Acanthis cannabina 228
Acanthis flammea 226
Acanthis flavirostris 228
Accipiter brevipes 323
Accipiter gentilis 66, 266
Accipiter nisus 66, 266
Acrocephalus arundinaceus 178, 317
Acrocephalus melanopogon 174
Acrocephalus paludicola 178
Acrocephalus palustris 180
Acrocephalus schoenobaenus 178
Acrocephalus scirpaceus 180
Aegithalos caudatus 208
Aegolius funereus 142
Aegypius monachus 58, 260
Aix galericulata 322
Alauda arvensis 156, 316
Alca torda 130, 315
Alcedo atthis 148
Alectoris barbara 324
Alectoris graeca 80
Alectoris rufa 80
Anas acuta 38, 248, 313
Anas americana 322
Anas angustirostris 322
Anas clypeata 40, 250
Anas crecca 40, 248
Anas penelope 42, 250
Anas platyrhynchos 38, 248, 313
Anas querquedula 40, 248
Anas strepera 38, 248
Anser albifrons 52, 256
Anser anser 52, 256
Anser caerulescens 323
Anser erythropus 52, 256
Anser fabalis 54, 256
Anthropoides virgo 324
Anthus campestris 164
Anthus cervinus 166
Anthus novaeseelandiae 328
Anthus pratensis 166
Anthus spinoletta 164
Anthus trivialis 166, 316
Apus apus 146
Apus melba 146
Apus pallidus 326
Aquila clanga 60, 262
Aquila chrysaëtos 60, 260
Aquila heliaca 60, 260
Aquila pomarina 62, 262
Aquila rapax 323
Ardea cinerea 30
Ardea purpurea 30
Ardeola ibis 322
Ardeola ralloides 32
Arenaria interpres 90, 272, 314
Asio flammeus 138
Asio otus 138, 316
Athene noctua 138
Aythya ferina 44, 250
Aythya fuligula 44, 252
Aythya marila 44, 252
Aythya nyroca 46, 250

Bombycilla garrulus 168
Botaurus stellaris 30
Branta bernicla 54, 258
Branta canadensis 323
Branta leucopsis 54, 258
Branta ruficollis 323
Bubo bubo 136
Bucephala clangula 42, 252
Bucephala islandica 322
Burhinus oedicnemus 114, 282, 315
Buteo buteo 64, 264
Buteo lagopus 64, 264
Buteo rufinus 64, 264

Calandrella cinerea 158
Calandrella rufescens 326
Calcarius lapponicus 222
Calidris alba 108, 282
Calidris alpina 108, 280
Calidris canutus 106, 280
Calidris ferruginea 108, 280
Calidris maritima 325
Calidris minuta 106, 280
Calidris temminckii 106, 280

Caprimulgus europaeus 144
Carduelis carduelis 226
Carduelis chloris 224, 319
Carduelis spinus 226, 319
Carpodacus erythrinus 230
Casarca ferruginea 323
Cepphus grylle 132, 315
Cercotrichas galactotes 328
Certhia brachydactyla 216
Certhia familiaris 216
Cettia cetti 174
Charadrius alexandrinus 92, 274
Charadrius dubius 92, 274
Charadrius hiaticula 92, 274, 314
Chlamydotis undulata 324
Chlidonias hybrida 124, 294
Chlidonias leucopterus 124, 294
Chlidonias niger 124, 294
Ciconia ciconia 36
Ciconia nigra 36
Cinclus cinclus 170, 317
Circaëtus gallicus 62, 262
Circus aeruginosus 68, 268, 313
Circus cyaneus 70, 268
Circus macrourus 70, 268
Circus pygargus 70, 268
Cisticola juncidis 174
Clamator glandarius 326
Clangula hyemalis 46, 252
Coccothraustes coccothraustes 222, 319
Columba livia 132
Columba oenas 132
Columba palumbus 134, 316
Coracias garrulus 146
Coturnix coturnix 82
Corvus corax 246
Corvus corone cornix 246
Corvus corone 246, 320
Corvus frugilegus 246
Corvus monedula 244
Crex crex 86
Cuculus canorus 144
Cursorius cursor 325
Cyanopica cyanus 242
Cygnus bewickii 56, 258
Cygnus cygnus 56, 258
Cygnus olor 56, 258, 313

Delichon urbica 160
Dendrocopos leucotos 154
Dendrocopos major 152
Dendrocopos medius 152
Dendrocopos minor 154
Dendrocopos syriacus 152
Dryocopus martius 150

Egretta alba 32
Egretta garzetta 32
Elanus caeruleus 323
Emberiza aureola 329
Emberiza bruniceps 328
Emberiza caesia 329
Emberiza calandra 218, 319
Emberiza cia 220
Emberiza cirlus 218
Emberiza citrinella 218, 319
Emberiza hortulana 220
Emberiza leucocephala 328
Emberiza melanocephala 222
Emberiza pusilla 329
Emberiza rustica 329
Emberiza schoeniclus 220
Eremophila alpestris 158
Erithacus rubecula 198, 318
Eudromias morinellus 94, 276

Falco biarmicus 323
Falco columbarius 74, 270
Falco cherrug 72, 270
Falco eleonorae 323
Falco naumanni 324
Falco peregrinus 72, 270
Falco rusticolus 324
Falco subbuteo 74, 270
Falco tinnunculus 74, 270, 313
Falco vespertinus 76, 270
Ficedula albicollis 194, 318
Ficedula hypoleuca 194
Ficedula parva 196
Francolinus francolinus 324
Fratercula arctica 130
Fringilla coelebs 224, 319
Fringilla montifringilla 224
Fulica atra 84, 314
Fulmarus glacialis 321

Galerida cristata 156
Galerida theklae 326
Gallinago gallinago 96, 314
Gallinago media 96
Gallinula chloropus 86, 314
Garrulus glandarius 240

Gavia adamsii 24
Gavia arctica 24
Gavia immer 24
Gavia stellata 24
Gelochelidon nilotica 126, 290
Glareola pratincola 112
Glaucidium passerinum 140
Grus grus 82
Gypaëtus barbatus 323
Gyps fulvus 58, 260

Haematopus ostralegus 90, 272
Haliaeëtus albicilla 58, 260
Hieraaëtus fasciatus 323
Hieraaëtus pennatus 62, 262
Himantopus himantopus 110, 272
Hippolais icterina 180, 317
Hippolais olivetorum 327
Hippolais pallida 182
Hippolais polyglotta 182
Hirundo daurica 162
Hirundo rupestris 160
Hirundo rustica 158, 316
Histrionicus histrionicus 322
Hydrobates pelagicus 321
Hydroprogne tschegrava 122, 290

Ixobrychus minutus 34, 313

Jynx torquilla 148, 316

Lagopus lagopus 76
Lagopus mutus 76
Lanius collurio 170, 317
Lanius excubitor 168
Lanius minor 168
Lanius nubicus 328
Lanius senator 170
Larus argentatus 118, 286
Larus audouini 325
Larus canus 120, 286
Larus fuscus 118, 286
Larus genei 325
Larus glaucoides 325
Larus hyperboreus 116, 286
Larus ichthyaëtus 325
Larus marinus 118, 286
Larus melanocephalus 120, 288
Larus minutus 122, 288
Larus ridibundus 120, 288, 315
Limicola falcinellus 325
Limosa lapponica 100
Limosa limosa 100, 315
Locustella fluviatilis 176
Locustella luscinioides 176
Locustella naevia 176
Loxia curvirostra 232
Loxia leucoptera 232
Loxia pytyopsittacus 232
Lullula arborea 156
Luscinia luscinia 200
Luscinia megarhynchos 200
Luscinia svecica 200
Lusciniola melanopogon 174
Lymnocryptes minimus 96
Lyrurus tetrix 78

Melanitta fusca 48, 254
Melanitta nigra 48, 254
Melanocorypha calandra 326
Melanocorypha leucoptera 326
Melanocorypha yeltoniensis 326
Mergus albellus 50, 254
Mergus merganser 50, 254
Mergus serrator 50, 254
Merops apiaster 148
Milvus migrans 68, 266
Milvus milvus 68, 266
Monticola saxatilis 202
Monticola solitarius 204
Montifringilla nivalis 234
Motacilla alba 164, 316
Motacilla cinerea 162
Motacilla flava 162, 316
Motacilla flava cinereocapilla 162
Motacilla flava feldegg 162
Motacilla flava flavissima 162
Motacilla flava thunbergi 162
Muscicapa striata 194

Neophron percnopterus 323
Netta rufina 42, 250
Nucifraga caryocatactes 242
Numenius arquata 98, 315
Numenius phaeopus 98
Numenius tenuirostris 324
Nyctea scandiaca 136
Nycticorax nycticorax 34

Oceanodroma castro 321
Oceanodroma leucorrhoa 321
Oenanthe hispanica 202

Oenanthe isabellina 327
Oenanthe leucura 327
Oenanthe oenanthe 202
Oenanthe pleschanka 327
Oriolus oriolus 240, 320
Otis tarda 84
Otis tetrax 84
Otus scops 140
Oxyura leucocephala 323

Pandion haliaëtus 72, 262
Panurus biarmicus 208
Parus ater 210
Parus caeruleus 212
Parus cinctus 326
Parus cristatus 210
Parus cyanus 326
Parus lugubris 326
Parus major 214, 318
Parus montanus 212
Parus palustris 212
Passer domesticus 236
Passer hispaniolensis 236
Passer montanus 236, 320
Pelecanus crispus 322
Pelecanus onocrotalus 322
Perdix perdix 80, 313
Perisoreus infaustus 240
Pernis apivorus 66, 264
Petronia petronia 234
Phalacrocorax aristotelis 322
Phalacrocorax carbo 28
Phalacrocorax pygmeus 322
Phalaropus fulicarius 112, 282
Phalaropus lobatus 112, 282
Phasianus colchicus 82, 314
Philomachus pugnax 110, 282, 315
Phoenicopterus ruber 322
Phoenicurus ochruros 198, 318
Phoenicurus phoenicurus 198, 318
Phylloscopus bonelli 190
Phylloscopus collybita 188, 317
Phylloscopus inornatus 328
Phylloscopus proregulus 328
Phylloscopus sibilatrix 190, 317
Phylloscopus trochiloides 192
Phylloscopus trochilus 190, 318
Pica pica 242, 320
Picoides tridactylus 154
Picus canus 150
Picus viridis 150
Pinicola enucleator 230
Platalea leucorodia 36
Plectrophenax nivalis 216
Plegadis falcinellus 34
Pluvialis apricaria 94, 274
Pluvialis squatarola 94, 274
Podiceps auritus 28
Podiceps cristatus 26, 313
Podiceps grisegena 26
Podiceps nigricollis 28
Podiceps ruficollis 26
Porphyrio porphyrio 324
Porzana parva 88, 314
Porzana porzana 88
Porzana pusilla 88
Prunella collaris 172
Prunella modularis 172, 317
Pterocles alchata 325
Pterocles orientalis 325
Puffinus griseus 321
Puffinus puffinus 321
Pyrrhocorax graculus 244
Pyrrhocorax pyrrhocorax 244
Pyrrhula pyrrhula 234, 319

Rallus aquaticus 86
Recurvirostra avosetta 110, 272
Regulus ignicapillus 192
Regulus regulus 192
Remiz pendulinus 210
Rhodopechys githaginea 328
Rhodostethia rosea 325
Riparia riparia 160, 316
Rissa tridactyla 122, 288

Saxicola rubetra 196
Saxicola torquata 196
Scolopax rusticola 98, 314
Serinus citrinella 228
Serinus serinus 230
Sitta europaea 214, 319
Sitta neumeyer 327
Sitta whiteheadi 327
Somateria mollissima 46, 252
Somateria spectabilis 322
Stercorarius longicaudus 116, 284
Stercorarius parasiticus 114, 284, 315
Stercorarius pomarinus 114, 284
Stercorarius skua 116, 284
Sterna albifrons 128, 290
Sterna dougallii 128, 292

Sterna hirundo 126, 292, 315
Sterna paradisaea 126, 292
Sterna sandvicensis 128, 290
Streptopelia decaocto 134
Streptopelia turtur 134
Strix aluco 140
Strix nebulosa 326
Strix uralensis 142
Sturnus roseus 238
Sturnus unicolor 238
Sturnus vulgaris 238
Sula bassana 321
Surnia ulula 142
Sylvia atricapilla 184
Sylvia borin 182
Sylvia cantillans 188
Sylvia communis 186
Sylvia conspicillata 328
Sylvia curruca 186, 317
Sylvia hortensis 184
Sylvia melanocephala 184
Sylvia nisoria 188
Sylvia rüppelli 328
Sylvia sarda 328
Sylvia undata 186

Tadorna tadorna 48, 256
Tarsiger cyanurus 327
Tetrao urogallus 78
Tetrastes bonasia 78
Tichodroma muraria 214
Tringa cinereus 325
Tringa erythropus 100, 276
Tringa flavipes 324
Tringa glareola 104, 278
Tringa hypoleucos 104, 278
Tringa nebularia 102, 278
Tringa ochropus 104, 278
Tringa stagnatilis 102, 276
Tringa totanus 102, 276, 315
Troglodytes troglodytes 172
Turdus iliacus 206
Turdus merula 208, 318
Turdus migratorius 327
Turdus naumanni 327
Turdus obscurus 327
Turdus philomelos 204
Turdus pilaris 206, 318
Turdus ruficollis 327
Turdus sibiricus 327
Turdus torquatus 206
Turdus viscivorus 204
Tyto alba 136

Upupa epops 144
Uria aalge 130

Vanellus gregarius 324
Vanellus spinosus 324
Vanellus vanellus 90, 272, 314

Xema sabini 325

Zoothera dauma 327